THE LEGITIMACY OF INTERNA

Global Environmental Governance

Series Editors: John J. Kirton, Munk Centre for International Studies, Trinity College, Canada and Miranda Schreurs, Freie Universität Berlin, Germany

Global Environmental Governance addresses the new generation of twenty-first century environmental problems and the challenges they pose for management and governance at the local, national, and global levels. Centred on the relationships among environmental change, economic forces, and political governance, the series explores the role of international institutions and instruments, national and sub-federal governments, private sector firms, scientists, and civil society, and provides a comprehensive body of progressive analyses on one of the world's most contentious international issues.

Titles in the series

Governing Agrobiodiversity
Plant Genetics and Developing Countries
Regine Andersen
ISBN 978-0-7546-4741-6

The Social Construction of Climate Change
Power, Knowledge, Norms, Discourses
Edited by
Mary E. Pettenger
ISBN 978-0-7546-4802-4

Governing Global Health
Challenge, Response, Innovation
Edited by
Andrew F. Cooper, John J. Kirton and Ted Schrecker
ISBN 978-0-7546-4873-4

Participation for Sustainability in Trade
Edited by
Sophie Thoyer and Benoît Martimort-Asso
ISBN 978-0-7546-4679-2

Bilateral Ecopolitics
Continuity and Change in Canadian-American Environmental Relations
Edited by
Philippe Le Prestre and Peter Stoett
ISBN 978-0-7546-4177-3

The Legitimacy of International Regimes

HELMUT BREITMEIER

Fernuniversität in Hagen, Germany

LONDON AND NEW YORK

First published 2008 by Ashgate Publishing

2 Park Square, Milton Park, Abingdon, Oxon OX14 4RN
711 Third Avenue, New York, NY 10017, USA

Routledge is an imprint of the Taylor & Francis Group, an informa business

First issued in paperback 2016

British Library Cataloguing in Publication Data
Breitmeier, Helmut
  The legitimacy of international regimes. - (Global
  environmental governance series)
  1. International agencies 2. Jurisdiction (International
  law) 3. Environmental policy - International cooperation
  I. Title
  341.2

Library of Congress Cataloging-in-Publication Data
Breitmeier, Helmut.
  The legitimacy of international regimes / by Helmut Breitmeier.
    p. cm. -- (Global environmental governance)
  Includes bibliographical references and index.
  ISBN 978-0-7546-4411-8 (alk. paper)
  1. Environmental policy--International cooperation. 2. Environmental protection--International cooperation. 3. International organization. 4. Legitimacy of governments. I. Title.
  JZ1324.B74 2008
  341.2--dc22
                                                                              2008030278

ISBN 978-0-7546-4411-8 (hbk)
ISBN 978-1-138-25911-9 (pbk)

Contents

List of Tables

Acknowledgements

This book has benefited from the intellectual guidance and the support of many people. Marc A. Levy, Oran R. Young, and Michael Zürn had initiated the project on the creation of the International Regimes Database (IRD). I joined this group soon after it had taken up its work. This took place in the context of a comprehensive research project on the Implementation and Effectiveness of International Environmental Commitments carried out at the International Institute for Applied Systems Analysis (IIASA) in Laxenburg (Austria). The development of a data-protocol that can be used for the coding of regimes remarked the first major achievement. The developers of this data-protocol benefited from the advice of members of the IIASA-project and of case-study experts who participated in trial runs carried out for the testing of previous versions of this codebook. After the end of an interim-period that followed the completion of the IIASA-project, new financial support was provided by the German Research Association (DFG) for the coding of a significant number of regimes. Additional funding for this coding phase became also available from the US National Science Foundation (NSF) and the German American Academic Council Foundation (GAAC). The discussions which I had with Oran and Michael during the coding phase expanded my horizon about the analysis of international institutions. I benefited substantially from their advice during the implementation of this project.

I am very grateful to the following experts who participated in the coding of 23 regimes: Matthew R. Auer, James S. Beckett, Thomas Bernauer, Pamela S. Chasek, Rene Coenen, Elizabeth Corell, Debbie Davenport, Leonard B. Dworsky, Peter Ehlers and members of the Helsinki-Commission, Christel Elvestad, David S. Favre, Robert L. Friedheim, Ray Gambell, Fred P. Gale, Andy Garner, Brian Hallman, Geir Honneland, Gudrun Henne, Richard A. Herr, James Joseph, Christopher C. Joyner, Jonathan P. Krueger, Jack Manno, Frank Marty, Britta Meinke, Radu Mihnea, Ronald B. Mitchell, Illya Natchkov, Kate O'Neill, Sebastian Oberthür, Edward A. Parson, Dwight Peck, Gerard Peet, MJ Peterson, Kal Raustiala, Lasse Ringius, Peter H. Sand, Sibel Sezer, Clare Shine, Jon Birger Skjaerseth, David G. Victor, Virginia Walsh, Jacob Werksman, Jorgen Wettestad, Koos Wieriks, Andrea Williams, and Bernhard Zangl. I admire their expertise about regime-related issues and the long-term experience which they gained as research scholars, participants in regime negotiations, leading staff members of international institutions and national ministries or agencies, or as representatives of non-state actors. I also owe a debt of gratitude to many students who gave advice on regimes that can be coded for the IRD or who helped to identify case-study experts that can take roles as coders. I would also like to thank the staff members of those 13 regime secretariats who filled out a separate questionnaire for the collection of information about rules that govern participation by various types of non-state actors in regimes.

During the process of data collection and the development of theoretical concepts that became established in this book, I strongly benefited from collaboration with my colleagues at the Department of Political Science at Darmstadt University of Technology. Klaus Dieter Wolf and his research group have been extremely supportive in various ways. Discussions with Klaus Dieter Wolf, Ingo Take and other colleagues in Darmstadt helped me to sharpen my view about regimes as arenas which are determined by conflict and interplay among various types of actors in world politics. Discussions with colleagues also caused me to reflect more critically on the influence of the broad range of non-state actors on the formation, evolution or effectiveness of international governance systems. Some of my thoughts about the impact of non-state actors on international institutions took shape in discussions with Volker Rittberger in Tübingen. During the creation of the IRD and the production of this book, I benefited substantially from the assistance of student assistants in Darmstadt. Nina Bressel, Andreas Haidvogl, Silke Hajunga, Gerhard Kilian, Thomas Kessler and Kemal Özden carried out a number of work-intensive tasks: the production of codebooks that were used for the coding of regimes; the review and editing of data delivered by coding experts; the production of additional queries addressed to coding experts pertaining to data which they had already delivered; the entering of data into the IRD. I am grateful to Markus Patberg and Gwendolyn Remmert for the editing of the manuscript. Further on, Joachim Hereth and Tim Kaiser from the Department of Mathematics in Darmstadt gave assistance when we exported the database from Oracle to MS Access. I am also grateful to Heike Jensen who dealt with much of the administrative tasks that inevitably occur in an international research project.

The development of case structures and the coding of regimes built on conceptual and theoretical reflections included in two working papers. They summarize the results of the work which the database-team completed during the development-phase at IIASA. The comprehensive data-protocol which includes all those variables and operationalizations that were used during the coding appeared as a co-authored working-paper by Breitmeier et al. 1996a. The conceptual thoughts about the design of the database and the criteria that govern the development of a case structure for the coding of a regime were described in another co-authored working paper by Breitmeier et al. 1996b.

List of Abbreviations

AKN	Aktionskonferenz Nordsee
AP	Actual Performance
ASOC	Antarctic and Southern Ocean Coalition
ATCM	Antarctic Treaty Consultative Meeting
ATCP	Antarctic Treaty Consultative Party
CCAMLR	Convention on the Conservation of Antarctic Marine Living Resources
CIESIN	Center for International Earth Science Information Network
CITES	Convention on International Trade in Endangered Species of Wild Fauna and Flora
CO	Collective Optimum
CPRs	Common-Pool Resources
CRAMRA	Convention on the Regulation of Antarctic Mineral Resource Activities
CSCE	Commission on Security and Cooperation in Europe
ECE	United Nations Economic Commission for Europe
ECOSOC	United Nations Economic and Social Council
EMEP	Co-operative Programme for the Monitoring and Evaluation of the Long-Range Transmissions of Air Pollutants in Europe
EU	European Union
FAO	Food and Agriculture Organization of the United Nations
FFA	Forum Fisheries Agency
GATT	General Agreement on Tariffs and Trade
GEF	Global Environment Facility
GLU	Great Lakes United
GNP	Gross National Product
GSP	Generalized System of Preferences
HELCOM	Baltic Marine Environment Protection Commission (Helsinki Commission)
IAATO	International Association of Antarctica Tour Operators
IAGLR	International Association for Great Lakes Research
IATTC	Inter-American Tropical Tuna Commission
ICBP	International Council of Bird Protection
ICCAT	International Commission for the Conservation of Atlantic Tunas

ICI	Industrial Chemical Industries
ICPR	International Commission for the Protection of the Rhine
ICRC	International Committee of the Red Cross
ICSU	International Council of Scientific Unions
IGOs	International Governmental Organizations
IIASA	International Institute for Applied Systems Analysis
IIED	International Institute for Environment and Development
IJC	International Joint Commission
IMO	International Maritime Organization
IMR	Norwegian Institute of Marine Research
IPCC	Intergovernmental Panel on Climate Change
IPED	International Panel of Experts on Desertification
IRD	International Regimes Database
ITTA	International Tropical Timber Agreement
IUCN	World Conservation Union
IWC	International Whaling Commission
IWRB	International Waterfowl Research Bureau
JCP	Baltic Sea Joint Comprehensive Environmental Action Programme
LRTAP Convention	Convention on Long-Range Transboundary Air Pollution
MARPOL Convention	International Convention for the Prevention of Pollution from Ships
MOU	Memorandum of Understanding
NAFTA	North American Free Trade Agreement
NASCO	North Atlantic Salmon Conservation Organization
NGOs	Non-Governmental Organizations
NO_x	Nitrogen Oxides
OECD	Organisation for Economic Cooperation and Development
OILPOL Convention	International Convention for the Prevention of Pollution of the Sea by Oil
OSPAR Convention	Convention for the Protection of the Marine Environment of the North-East Atlantic
PINRO	Russian Polar Research Institute of Fishery and Oceanography
RAP	Remedial Action Plan
SCAR	Scientific Committee on Antarctic Research
SIRs	Systems for Implementation Review
TFACT	Trust Fund for Assistance to Countries in Transition
TRAFFIC	Trade Record Analysis of Flora and Fauna in Commerce
UNCCD	United Nations Convention to Combat Desertification
UNCLOS	United Nations Convention on the Law of the Sea
UNCTAD	United Nations Conference on Trade and Development
UNEP	United Nations Environment Program

UNFCCC	United Nations Framework Convention on Climate Change
USA	United States of America
USSR	United Socialist Soviet Republic
WBCSD	World Business Council on Sustainable Development
WCMC	World Conservation Monitoring Centre
WHO	World Health Organization
WIPO	World Intellectual Property Organization
WTO	World Trade Organization
WWF	World Wide Fund for Nature (formerly World Wildlife Fund)

Chapter 1

Introduction

How can social order beyond the level of the nation-state emerge, be maintained, or further develop under conditions of profound change? The changes occurring in world politics in the past few decades require a new mix between coercion, self-interest, or legitimacy as the possible modes by which order can be established beyond the nation-state. These three modes of social order are not fully congruent with the categorization of power, interest, or knowledge as variables which are usually used for explaining outcomes and behavioral changes in world politics (Hasenclever, Mayer and Rittberger 1997; Levy, Young and Zürn 1995; Young and Osherenko 1993). This traditional categorization, however, seems no longer justified as far as the mode of knowledge is concerned. The more the character of politics in multi-level governance systems approximates, even though not fully equals, to the pluralist character of domestic politics, the less it remains plausible that studies of domestic or international order rely on partly different understandings with knowledge considered to become relevant on the international level and legitimacy to unfold mainly in domestic politics. Further on, the more international law will constrain the autonomy of political decision-making within the nation-state or will permeate the lifeworld of the people on the domestic level, the more will it become necessary to establish grounds based on social reason which justify to follow or implement international legal norms.[1]

Though this study focuses on legitimacy and will move away from the traditional understanding of knowledge as a basic causal variable that accounts for social order, many insights gained from knowledge-based theories cannot simply be abandoned in searching for components which make up legitimacy as a concept. The various knowledge-based approaches on international institutions were subsumed by Andreas Hasenclever, Peter Mayer and Volker Rittberger (1997, 136–210) under the labels of "weak" and "strong" cognitivism. They called these approaches "cognitive" because these theories prefer an interpretive to the traditional behavioral approach to politics. Cognitive theories explain outcomes and behavioral change as a result of inter-subjective meanings and shared understandings. How far can we tie in cognitive theories in developing a conception of legitimacy that applies

---

1   See also Hurd (1999, 380) who argues that the mode of legitimacy has been neglected by the scientific research of international politics. A growing number of studies produced by political scientists or lawyers which evaluate the legitimate character, or reflect on the prospects of improvements of the legitimacy of governance on the global level or within regional institutions like the European Union signals that such neglect is no longer tenable (Bodansky 1999; Cutler 2001; Habermas 1998; Wolf 1997).

to global governance systems? The theories subsumed under *weak cognitivism* depict a narrow part of the real world of international politics as they consider knowledge to become relevant mainly for the re-evaluation of state interests. They highlight the epistemic foundations on which grounds for legitimacy must be based. These approaches take into account that consensual knowledge produced by networks of scientific and technical experts about cause-effect-relationships sets the benchmark for assessing the quality of policies. They lead us to ask how far global governance is capable of learning or can absorb new ideas and adapt to new requirements arising from constant change in world politics (Adler and P. M. Haas 1992; Goldstein and Keohane 1993; E. B. Haas 1990; P. M. Haas 1990; Mitchell, Clark, Cash and Dickson 2006). However, weak cognitivist theories inhere an elitist approach of international policymaking. They mainly describe the collaboration between, or the strategies taken by, scientific and technical experts or epistemic networks which facilitate change in the minds of policymakers.

The theories merged under the label of *strong cognitivism* take the assumption as a starting-point that the normative environment of international politics in which states became embedded shape the identities of, or contribute to socialize, states in international society. They also allude to the potential inherent to globalized public spheres for establishing a discursive setting through which an integrating global civil society can communicate, become aware of its collective will and reach consensus on the rightness of arguments and the appropriateness of international policies (Bohman 1998; Risse 2000). Unlike weak cognitivist approaches, these theories are much more able to cope with the new reality of an emerging global civil society influencing political decision-making on the level beyond the nation-state. The distinction between weak and strong cognitivism offers a way to rearrange the multitude of cognitivist theories under the umbrella of two broader strands of literature. But it is questionable whether the original meaning of knowledge applies to all of the theories which were labeled as "cognitivist" equally.[2] Strong cognitivism considers the interplay between international norms and evolving pluralism in world politics. It focuses on cognition and communication as processes that determine perceptions and preferences by states. But it ignores the traditional meaning of the knowledge as a relevant causal factor.

This book deals with the following questions: How far do the policies and procedures of international regimes justify obedience in world politics? Did institutional mechanisms contribute to improving the legitimacy of international institutions? How far can non-state actors have an effect on the legitimacy of international environmental regimes? These questions extensively preoccupied students of international politics and political decision-makers for a number of reasons. The internationalization of policymaking described by international

---

2   For example, Ernst B. Haas (1990, 21) defined consensual knowledge as "generally accepted understandings about cause-effect-linkages about any set of phenomena considered important by society, provided only that the finality of the accepted chain of causation is subject to continuous testing and examination through adversary procedures".

lawyers, political scientists or sociologists under the terms of global juridification or legalization suggests the rise of new systems of political rule on the transnational level. Policymaking on the domestic level has been confronted with a number of questions that apply to international institutions likewise: Which benefit arises from governance systems? Would the state of the world be better or worse without institutions? How far are outcomes and impacts produced by institutions acceptable to social groups or states? Do global governance systems consider the demand for transnational democracy?

Within the past few decades world politics has experienced a strong increase of governance systems for the protection of the environment. The complexity of global governance systems and their interconnection with domestic politics imply that obedience to global governance can be realized only if a transition from hierarchical or state-dominated forms of international governance towards horizontal forms of global governance will be made. Such horizontal forms consider the potential which global civil society can provide for the achievement of outputs, outcomes, and impacts in world politics. Many approaches dealing with emerging global civil society or with different types of non-state actors in world politics implicitly or explicitly demand such a transition. The prime hypothesis which will be followed in this study draws a causal connection between the activities of non-state actors and legitimacy: Only if non-state actors can develop political activities and give effect to their problem-solving potential, it will be possible to improve the legitimacy of global governance systems. This causal connection is now commonly accepted by policymakers and appears as a commonplace in the declarations adopted by intergovernmental conferences. But we still lack knowledge whether this causal assumption is confirmed by the reality. We also lack a common understanding about the concept of legitimacy: Why can social order in world politics emerge if global governance systems fulfill grounds for legitimacy? What does the idea of legitimacy mean? Do non-state actors have an impact on outcomes and impacts in world politics? These questions will comprehensively be dealt with in the different chapters of this book.

In the following, the term of non-state actors will be used as a conceptual framework for exploring the influence of various types of actors other than the state. This implies that attention will not be limited to non-profit activist interest groups, but analysis will consider a broad spectrum of non-state actors. This study is not biased in favor of a single type of non-state actor. It takes into account that transnational politics is characterized by heterogeneous interests of different types of non-state actors. This study is also aware that non-state actors alone can hardly solve all the complex issues that exist in world politics. The political monopoly of the international society of states has eroded since World War II, but the state continues to play an important role in global governance systems. One should not be blind to the problems that can arise from strengthened participation of non-state actors in political processes. The empirical chapters of this book will illustrate how far non-state actors can influence processes and outcomes of global governance.

The International Regimes Database

The prime hypothesis will be confronted with empirical findings which can illustrate the impact of non-state actors on the legitimacy of global governance. Most of the empirical parts of this study will be based on the results of the International Regimes Database (IRD). Twenty-three international environmental regimes have been coded for the IRD (Breitmeier, Young and Zürn 2006). This database emerged in the context of a multi-year collaborative project between American and German scholars (Breitmeier, Levy, Young and Zürn 1996a and 1996b). This project was carried out in two stages. In a first step, a comprehensive data protocol was developed by Marc A. Levy, Oran R. Young, Michael Zürn and myself when we were involved with the so-called IEC-project at IIASA in Laxenburg (Austria). When the coding of 23 international environmental regimes began in fall 1998, the project entered into a second stage. The coding of these regimes has been carried out by international case study experts. Well-defined criteria have been used for dividing these regimes into 92 sub-cases. These criteria have been used to distinguish so-called "regime elements" as smallest units of analysis. Regime elements were identified with coding case study experts if the existence of separate institutional arrangements (e.g., conventions, protocols, annexes, or relevant soft law agreements) allowed to separate a regime into various components. These criteria also provide for the distinction of different periods in the lifecycles of regimes. They were used for identifying important watersheds (e.g., the moment of the development of new institutional arrangements) in the lifecycles of regimes. Fourteen of the 23 regimes coded came into existence more than two decades ago. The IRD provides a tool for analyzing regimes from a historical-comparative perspective. This tool became publicly available to the social science community (see Breitmeier, Young and Zürn 2006). Empirical analysis will also use another less comprehensive data set which was collected independently of the IRD. This separate data-set will be used for exploring the procedures provided by environmental regimes for participation of non-state actors. The collection of these separate data on procedures for participation became necessary when this individual research project about the impact of non-state actors on global governance has been launched after the development of the IRD-codebook had been concluded. The separate data-set on procedures for participation will be used for empirical analysis in chapter 10, but it is less relevant to analysis carried out in other empirical chapters.

Outline of the Book

Theoretical and empirical chapters balance one another in this book. Four theoretical and conceptual chapters will illustrate the relevance and meaning of legitimacy, the impact of institutional mechanisms and non-state actors on environmental governance, or describe various methodological issues involved with the coding of 23 environmental regimes. Chapter 2 explores which consequences arise from changes in world politics for the evolution of social order beyond the nation-state. It will be argued that change in world politics needs to strengthen legitimacy as a mode of social order in world society. The dependent variable of legitimacy will be operationalized in chapter 3. The concept of legitimacy used in this study will be differentiated from Max Weber's traditional concept of legitimacy who understood the subject in terms of people's belief in legitimacy. Instead, legitimacy will be understood as a concept that is based on grounds that justify obedience to social order in world politics. These grounds are connected to the fulfillment of various outputs, outcomes and impacts of international regimes or of normative requirements that constitute politics in the modern era. The concept of the legitimacy of international regimes is based on five grounds that justify obedience. These grounds involve that obedience to the policies of international institutions will be justified i) if they improve our knowledge about the causes and effects of problems and about the policy options which can be used for problem-solving, ii) if they strengthen compliance, iii) if they change the state of the world in issue-areas towards a more desirable direction, iv) if they contribute that the distribution of costs and benefits will become acceptable to affected social groups or states, and v) if they provide procedures which allow participation by emerging civil society. The independent variable which focuses on possible contributions of non-state actors will be operationalized in chapter 4. This chapter distinguishes between various types of non-state actors and illustrates their activities and conflicting interests in international regimes. It will become obvious that non-state actors can promote or constrain the achievement of grounds for legitimacy in world politics. The question concerning the contribution of non-state actors to the fulfillment of various grounds for legitimacy will be answered in two different steps. First, it will be explored whether the international institution had an effect and whether institutional mechanisms existed which could be used by non-state actors for influencing the legitimacy of regimes. A second explanation considers participation by non-state actors in political processes. It will be explored how far non-state actors affected outcomes and impacts in institutional mechanisms or shaped discussions in transnational public spheres. Chapter 5 builds the bridge between theoretical chapters and empirical analysis. It describes the guidelines used, and it illustrates the experiences made, during the development of case structures for the coding of regimes.

In the five empirical chapters that follow it will be explored how far the prime hypothesis is confirmed by empirical reality. Chapters 6–10 will present results of empirical analysis. It will be explored in these chapters how far various grounds for legitimacy have been fulfilled. It will be explored whether institutional mechanisms influenced observed outputs, outcomes and impacts and whether non-state actors contributed to achieving observed levels of change in our dependent variable. Chapter 6 will explore whether a reduction of uncertainties in the cognitive setting took place in regimes. It will be measured whether regimes were affected by change regarding the understanding about the nature of the problem, the completeness of information about policy options, or the capacities which enable members to participate in social practices. Chapter 7 will illustrate how far compliance as another ground for legitimacy has been achieved in regimes. Two different measurements will be made. It will be measured how far compliance has been achieved on the level of all members of single regimes. Another measurement directs attention on compliance behavior of single important states in a regime. Chapter 8 uses two different approaches for measuring changes in the state of problems managed by environmental regimes. The study of goal-attainment focuses on exploring whether goals which were established in the constitutive legal agreements of regimes have been fulfilled. The study of problem-solving directs the view on the measurement of the improvement, worsening, or continuation of the status quo in the state of environmental problems. Chapter 9 pays attention to the distributional consequences of regimes. Analysis will focus on patterns that determined the financing of the management or administration of regimes, and on the distribution of costs or benefits in issue-areas. Finally, chapter 10 explores whether a number of basic rights which characterize our concept of transnational democracy are provided in the rules of procedures of regimes and whether these rights were used by non-state actors. In some respects, this chapter has an explorative character since the total of regimes covered is less comprehensive than in other empirical chapters.

Chapter 2

International Regimes in a World of Change: Why Legitimacy?

What do we understand by a social order at the level beyond the nation-state? The term of a social order refers to the regularity, predictability, or stability of patterns of behavior in social life. The quality of social order can be determined by assessing whether governance systems lead to the fulfillment of specific goals underlying a social order. Hedley Bull (1977, 4) pointed out that social order "is not *any* pattern or regularity in the relations of human individuals or groups, but a pattern that leads to a particular result, an arrangement of social life such that it promotes certain goals or values" [*emphasis by author*]. Social order is desirable only if the goals which it fulfills are compatible with social reason. The different international orders that emerged in the past few centuries considered the state as a main actor which dominates interactions and the creation of outcomes in international politics.[1] The idea of the state as a main actor in international politics emerged with the Westphalian model of order. This type of international order was established after the political chaos and belligerency which had characterized relations among various types of actors in times before and during the Thirty Years' War in Europe.

The Westphalian model of order involved a sharp distinction between *internal* and *external* levels of policymaking. The principle of state autonomy underlying this order gave states complete authority within own national borders. It produced two separate political spaces on domestic and international levels. The number of goals pursued in governance systems grew strongly on both levels during the past centuries. This led to a gradual change in the traditional mix between the three modes of power, interest, and legitimacy. At first this change took place to a much stronger degree domestically than internationally. Security as the originally main goal of governance on both levels of policymaking was complemented by the emergence of other goals. As a result, orders which predominantly relied on power had to shift to the use of alternative modes of governance.

The number of goals began to increase on the domestic level and was accompanied by a redefinition of the role of the state and by a broad expansion of state activities. The modern state took shape in the eighteenth century. It suffered

---

1 The majority of contemporary theories of international politics still share this assumption of state-centrism defining political order "primarily in terms of negotiated connections among externally autonomous and internally integrated sovereigns" (March and Olsen 1998, 945).

from strongly grown costs of warfare. This was the reason why costly armies of mercenaries had to be replaced by standing armies. The expansion of state authority resulting from the reorganization of the military was taken as an opportunity by the people to demand of the state fulfillment of goals other than external security. The state became considerably engaged in the construction of domestic infrastructure or in social and economic affairs. State bureaucracies and new forms of taxation expanded. Governmental institutions increasingly affected social life. But political systems were also confronted with claims of the people for safeguarding civil rights against the arbitrary use of power. The past centuries were characterized by emancipation of domestic societies from different forms of authoritarian rule in many (though not all) countries of the world. On the level beyond the nation-state, demands for democratic legitimization have been expressed only more recently.

Political goals proliferated not only on the domestic but also on the international level. National societies considered their governments as managers of transnational problems. Many political problems could no longer be resolved by national self-help-strategies, but they required collective action by states. In the following, two questions will be addressed that deal with the relevance of legitimacy as a mode of social order: How far were past international orders based on legitimacy? Which reasons support the assumption that an institution-based order in world politics will have to rely more on the mode of legitimacy? It will become obvious that international institutions can produce partly negative consequences for national polities or for democratic decision-making at the domestic level. This leads to the conclusion that social order must in fact more be built on legitimacy. Moreover, transnational societization involves that international regimes must adapt to the participatory claims of emerging global civil society.

Global Juridification: Power and Self-Interest as Original Driving Forces

International regimes are elements of an institution-based international order. The evolution of this order began in the nineteenth century and was continued in the inter-war period within the League of Nations. After the end of World War II, this order began to take on a new quality. The scope of issues managed by international institutions has significantly expanded. The density and specificity of norms and rules that were established in these governance systems have increased likewise. The creation of international regimes expressed nothing less than a process of global juridification (Shapiro 1993; Slaughter 1995). The average yearly number of newly founded international environmental treaties amounted less than one between 1900 and 1945. This number rose to about nine per annum since 1960 (Frank, Hironaka and Schofer 2000, 100).[2]

---

2 A survey produced by the Bremen globalization-project also suggests that a significant growth of international environmental legal institutions emerged since the 1960s. The findings of this project support the general conclusion that juridification in

Legitimacy as a variable that accounts for order led a shadowy existence in international politics. This level of policymaking was dominated by coercive power and self-interest throughout the modern era (Hurd 1999). The evolution of an institution-based order in international politics was not exclusively based on coercive power. It also reflected the interest of an international community of states to realize security or other goals. Hedley Bull (1977, 67) argued that the "maintenance of order in international society has as its starting point the development among states of a sense of common interests in the elementary goals of social life". International regimes were long mainly perceived of as instruments for the production of outcomes and behavioral changes in world politics. They were considered as variables which intervene between basic causal conditions like power and self-interest on the one hand and changes in outcomes and behavior on the other (Krasner 1983, 5–10). Legitimacy has long played a marginal role in international politics, but it has never fully disappeared.[3] More than three decades ago, Inis L. Claude (1966) indicated that legitimacy becomes relevant also in international politics. The author argued that statesmen intend to justify their national foreign policies not only to themselves or their peoples. They also intend to earn the seal of approval from the United Nations or other international institutions and to avoid the stigma of disapproval by these institutions. This traditional conception of legitimacy is closely linked with a state-centric understanding of international politics. It considers legitimacy as a matter of intergovernmental relations. The author himself admitted that judgments of the United Nations about legitimacy "represent the preponderant opinion of the foreign offices and other participants in the management of the foreign affairs of the governments of Member States" (Claude 1966, 372). The study of international regimes long focused on exploring the role of power and self-interest for the evolution of institution-based international order. Two research projects which were completed in the early 1990s informed about the relevance of several causal factors for the formation of international regimes. A collaborative research project at Dartmouth College used five environmental regimes for the testing of power-, interest-, knowledge-based and contextual hypotheses (Young and Osherenko 1993). The Tübingen research group explored 13 security, economic, environmental, and human rights regimes that were established in East-West- or West-West-relations. The Tübingen group used these cases for the testing of power-, problem-, situation-structural, and other systemic hypotheses (Rittberger and Zürn 1990; Efinger, Mayer and Schwarzer 1993).

---

international politics took shape in the second half of the twentieth century, but it unfolded mainly in the OECD-world. This process occurred to a lesser extent in the developing world (Beisheim, Dreher, Walter, Zangl and Zürn 1999, 327–56).

3　While power was associated with self-interest, these two modes were occasionally completed by fulfillment of concerns regarding the legitimate character of international order. In a study which investigates the prerequisites to the resolution of international war between 1648 and 1989 Kalevi J. Holsti (1991, 241) concludes that "those peace settlements that were not considered legitimate by the vanquished and other states were soon threatened or overthrown".

Which findings emerged from the two projects? Firstly, both projects disprove the validity of a predominantly power-based understanding of international order. They disconfirm the pure form of hegemonic stability theory. While they acknowledge that other types of power have some relevance, they also reveal the limitations of power-based explanations.[4] Secondly, both projects confirm the validity of their own interest-based approaches of regime formation.[5] The Tübingen project concluded that its situation-structural hypothesis after which regime formation will i) most likely occur in situations which equal the co-ordination game type, ii) be possible in situations which come under the dilemma-type only if exogenous factors exert a favorable influence, or iii) be impossible in situations which come under the Rambo game type "clearly holds more promise than others of serving as a basis for successfully predicting and thus explaining regime formation" (Efinger, Mayer and Schwarzer 1993, 269). The Dartmouth project concluded that among the four central ideas (including power, interests, knowledge, and context) tested, "the case studies consistently emphasized factors affecting interest-based behavior in the context of interactive decision-making" (Young and Osherenko 1993, 230). Thirdly, findings of both projects suggest that power and self-interest are not mutually exclusive. These projects lead us to conclude that the evolution of institutional order results from a more or less unbalanced mix between these modes. However, the findings reveal that the creation of international regimes is motivated by utilitarian attitudes of states and that the use of various forms of power can increase the likelihood of institution-building.

At first sight, these findings contradict the argument that legitimacy will become more relevant in international politics. If international regimes are understood as instruments of change, the utility expected to arise from these institutions provides a reason that justifies institution-building. In this regard, self-interest and legitimacy coalesce. States can be considered to be interested in institution-building and to comply with the rules of international regimes since they regard them as legitimate as mechanisms for the attainment of common goals which can not be realized otherwise. But the relevance of interest- and power-based approaches decreases if analysis is no longer carried out on the basis of a state-centric understanding of

---

4 Under the circumstances, propositions made by hard-core power theorists of international order who, like John A. Hall (1996, 168), maintain that "that such regimes remain the subjects of American power, for all that their presence sometimes masks the operations of that power" do often not correspond with the empirical reality of late twentieth century international politics.

5 Mark W. Zacher (1996) came to a similar finding in exploring the long-term development of international shipping, air transport, telecommunications, or postal regimes which originated in international treaties or bodies that were established in the second half of the nineteenth or in the first half of the twentieth century. His study diagnoses that the norms of these regimes have been stable over approximately a century although the distribution of power capabilities between members of these regimes has been subject to dramatic changes. Consequently, he concludes that "mutual interests among states provide the bases for the development of many important international regimes" (Zacher 1996, 224).

international politics. This traditional concept neglected that emerging global civil society will become relevant as a factor for the evolution of social order in world politics and that the willingness to obey to international norms and rules depends on the agreement of transnational civil society. While interest-based theories of international regimes were developed in the context of a more general interest that evolved in the social sciences concerning the value, working, or change of institutions in the modern world, they were soon confronted with theoretical reservations that applied to the analysis of policymaking on both domestic and international levels.[6]

Expanding the Scope of Law: From International to World Society

The understanding of international regimes as elements of a process of global juridification coincides with the conception of international society developed by the English School. This theoretical approach argues in the tradition of Hugo Grotius that international law fulfills the functions of codifying the basic rules of coexistence, regulating international cooperation and placing constraints on international conflict. Hedley Bull (1977, 13) described the evolution of an international society of states as a process where states "conceive themselves to be bound by a common set of rules in their relations with one another, and share in the working of common institutions".[7] The model of international society is based on the assumption that international anarchy is a prevailing condition that must be considered in establishing order on that level. But the model of international society refuses the assumption that the character of relations between states is analogous to the Hobbesian notion of a state of nature between individuals on the domestic level. This Hobbesian notion implies that individuals are capable of orderly social life only under the rule of governmental power. The model of international society describes an order which no longer depends exclusively on the power capabilities of single states. This model emphasizes that order can emerge from the readiness of states to follow international law even if international norms will, in particular cases, contradict national interests. Tony Evans and Peter Wilson (1992, 335–6) pointed out that there are also important differences between the English School and

---

6   The mini-revolt against the relevance of rational choice in the study of domestic political processes and "against the self-interested model of the way a democratic polity actually works" observed by Jane J. Mansbridge (1990, 19) also spread to the study of international institutions. For example, Robert O. Keohane (2001, 6) admitted that rationalist theory "often carries with it the heavy baggage of egoism" and that rationalist theorists "recognize that the assumption of egoism oversimplifies social reality".

7   Barry Buzan (1993, 333–6) characterized Bull's model as a functional rather than civilizational conception of international society. The civilizational (or *gemeinschaft*) understanding sees society as an organic and traditional phenomenon which includes bonds of common sentiment, experience, and identity and which evolves in a historic process. The functional (or *gesellschaft*) understanding considers the evolution of a society as an act of will which finds expression in contracts consciously made.

regime theory. Regime theory starts from an issue-area-approach of international politics whereas the English School describes a universal international society. While such conceptual differences certainly exist between both theories, it can be argued that international society provides the legal framework (e.g., pacta sunt servanda) that guarantees the functioning of international regimes.

The expansion of international law arose from the insight that globalization can limit effective governance on the domestic level. In the era of globalization effective governance can only be accomplished if states will coordinate their policies and collaborate. The internationalization of policymaking was often in the interest of the national peoples. Therefore international institutions were long primarily analyzed as to whether they contribute to the attainment of various goals or to problem-solving. The first wave of studies on regime effectiveness focused on effects of international regimes which influence the general framework for the political management of trans-boundary problems. Regimes were understood as mechanisms which produce effects that were labeled as the *3 C's* (Levy, Keohane and Haas 1993, 398–408), because they raise *concern* about cause-effect relationships, improve the *contractual* environment and thereby foster the establishment of negotiation processes or reduce transaction costs in issue areas, or enhance *capacity*-building in developing countries. This strand of effectiveness research avoided to draw a direct causal connection between described effects of regimes and their impact on problem-solving or on other possible dimensions of effectiveness.

Subsequent studies have put more attention on the causal relationship existing between the performance of institutions and the different outcomes and impacts produced by them in single issue areas. These studies improved our understanding about the range of possible effects which must be considered in assessing whether the regulations resulting from the operation of regimes will deserve obedience by states and global civil society. The various approaches involved with the study of regime effectiveness were illustrated by Oran R. Young and Marc A. Levy (1999, 3–6):[8] Firstly, the *problem-solving approach* implies to determine the degree to which regimes will contribute to change the problem(s) in the issue area concerned; secondly, the *legal approach* involves to measure the degree to which contractual obligations are met; third, the *economic approach* adds an efficiency criterion to legal effectiveness because it evaluates the cost efficiency of policies initiated by regimes; fourthly, the *normative approach* evaluates whether normative principles such as justice, or participation are fulfilled by a regime; finally, a *political approach* perceives of problems as "functions of specific constellations of actors, interests, and institutions" (Young and Levy 1999, 5) and explores whether regimes can contribute to the management of the targeted problem by causing changes in the behavior of actors, by changing the interests of actors, or by initiating changes in regime-related policies. Some of these dimensions of regime effectiveness are closely related with the need for utility-maximization, whereas other dimensions derive grounds for obedience from normative conceptions.

---

8   See also Young (1994, 140–60) and Young (1999b, 108–25).

Whether international institutions can be used as instruments for the achievement of goals was very much disputed among researchers of international politics (Strange 1983; Mearsheimer 1995). Neo-realists and neo-institutionalists discussed the question about the causal relevance of international institutions in terms of whether regimes matter or not. But it was not considered in this debate that the answer to this question can also lie somewhere between these two extremes. The reality that can be found in international politics demonstrates that answers to the question on whether regimes contribute to problem-solving elude such a binary conception. A more diversified measure scale is required that embraces the different degrees of effectiveness. The antagonism which characterized the debate between neo-realists and neo-institutionalists prevented that these schools became aware of the full spectrum which exists in regard to the production of outcomes and impacts by international regimes.

Findings about regime effectiveness that were produced on the basis of the case-study-approach were often incomparable either because they varied in how they perceived of their dependent variable or because they used different independent variables for explaining regime effectiveness. The study of international regimes led to the diversification of knowledge, but it also suffered a lack in the accumulation of knowledge. The variety of dependent and independent variables used by the large number of empirical case studies about regimes made it very difficult to generalize theoretical findings. Already a decade ago, regime analysts diagnosed that "further advances in empirically based and theoretically oriented research on international regimes would considerably benefit from a *data base* built up by interested researchers following widely agreed-upon guidelines" (Mayer, Rittberger and Zürn 1993, 429). Quantitative analysis needs to spend an amount of resources in the creation of an International Regimes Database (IRD) that goes much beyond the resources which must be invested in the production of single case studies. But such a tool can make a substantial contribution to generalize findings about the causal impact of international regimes on behavioral changes and outcomes (Breitmeier, Levy, Young and Zürn 1996a and 1996b).

The model of international society considers global juridification as a process that concerns predominantly the world of states. The evolution of social order on the international level is considered mainly as a result of self-interest which motivates states to collaborate. This model disregards consequences that can arise from global juridification on domestic or transnational levels. By expanding the analysis of international regimes beyond the limits of state-centrism and beyond the scope of inter-state law, it will be possible to direct our view on sociological developments that occur on the level of emerging global civil society. These actors increasingly influence the evolution of international law. Global civil society is an important factor for achieving obedience to international law. Therefore, regimes will also be explored as arenas which are affected by transnational societization. This process involves that regimes are increasingly confronted with demands to justify outputs, outcomes and impacts or the procedures used for collective decision-making vis-à-vis emerging global civil society. The transformation of Westphalian order changed the general framework within which regime politics takes place. International regimes are now targets of change. Different types of actors influence decision-making or contribute to the implementation of regime policies.

Regime Consequences on Domestic and Transnational Levels

Which consequences of international regimes lead us to conclude that self-interest alone can hardly produce social order in world politics? Two types of effects on the democratic process have been ascribed to international regimes. Firstly, various regimes that were established under the umbrella of the United Nations indirectly fostered processes of democratization and contributed to the protection of human rights in many countries. Secondly, democratic theorists and students of global governance increasingly focused on analyzing the possible de-democratizing effects of international governance systems on national democracies (Dahl 1994; Held 1997; Wolf 2000). These two different views create a dilemma for those who analyze international institutions, since they emphasize the causal role of international regimes for both *democratization* as well as *de-democratization* of national polities (Breitmeier 2004).

A possible interpretation for these different findings involves that the effects of regimes on democracy depend on the context. Democratizing effects caused by international regimes can unfold particularly in those countries where civil society still lives under authoritarian rule. The thesis about possible de-democratizing effects on national polity can be considered to apply more to those liberal-democratic states in the Western hemisphere which are embedded in dense webs of international institutions. The existence of international norms for the protection of human rights forced many authoritarian states into political discourses on domestic and international levels where they had to justify their behavior to the international society of states and to their own societies. The studies which are included in a volume edited by Thomas Risse, Stephen C. Ropp, and Kathryn Sikkink (1999) intend to explain why some human rights norms that are established in global human rights agreements were internalized and implemented domestically. Non-governmental actors were ascribed a crucial role in this process of norm internalization since they acted as carriers and translators of these norms. Non-governmental actors often lack possibilities in authoritarian states to pressurize their governments with the aim of improving the human rights situation. But they can establish transnational alliances with NGOs abroad and politicize the human rights situation on the transnational level. This "boomerang effect" makes it possible for transnational human rights networks to put pressure on a norm-violating state. It can increase the legitimacy of this protest because it demands fulfillment of international norms on the domestic level (Keck and Sikkink 1998, 12–13).

Global juridification implies that analysis can no longer be confined to the character of regimes as an intervening variable between basic causal factors and outcomes. International institutions must also be understood as an *independent variable.* Oran R. Young (2002a) illustrates that it is often impossible to distinguish whether causation occurs on the basis of international environmental regimes as intervening or independent variables. For example, adherents of interest-based explanations admitted that interests are not only exogenously given, and that it will not suffice that regimes are considered as mechanisms for the translation of individual preferences into collective outcomes. Regimes can lead to a redefinition of interests with regime members. A similar argument can be made with respect to the treatment of knowledge as a possible basic causal factor. Adherents to cognitivism assume that knowledge determines outcomes and impacts because regime policies are adapted to the changing consensual knowledge. But causation can also occur in the reverse direction when a regime contributes to changing discourse or the quality of knowledge. Therefore, Young (2002a, 7) concludes that a causal chain whether it will understand regimes primarily as intervening or independent variables will inevitably "trigger a regress that has no natural stopping point".

Which implications arise from this dilemma for our study? The density of institutionalization has significantly grown in some contexts of world politic, but especially among industrialized countries. These strongly institutionalized contexts can interfere in social life very deeply. It is doubtful whether variables like power and self-interest, which have influenced efforts by states to establish international institutions, can still exclusively determine policymaking in international regimes. If one regards international regimes as an independent variable, analysis must also consider that institutions can partly elude the instrumental character intended by states. States and their societies realize that international institutions can become more independent from international society than their founders may originally have expected. Several consequences of global juridification occurring on domestic and transnational levels lead us to conclude that there is a growing need for legitimacy in world politics.

First, the internationalization of politics seems to strengthen de-parliamentarization at the national level. The internationalization of policymaking seems to strengthen governments vis-à-vis their societies and democratic political institutions. Multi-level negotiation systems allow governments to negotiate and sign international agreements before domestic democratic institutions will be asked for approval. Governments can blackmail parliaments by pointing out that rejection of the international agreement concluded by the government will endanger the establishment of the regime as a whole or damage the international reputation and credibility of the state. Though de-parliamentarization may unfold as a problem particularly in the European Union, it is hardly limited to the context of European institution-building.[9] Admittedly, de-parliamentarization can occur also independently of the effects of international governance. But de-parliamentarization is also caused by the internationalization of policy-making,

---

9   von Beyme (1997, 185–6) illustrates that a growing number of national laws passed by the German parliament in the past few decades has been caused by the European Union and international institutions.

because multi-level policymaking gives governments more latitude for realizing their political goals. Under these circumstances, it became more difficult for national parliaments to influence intergovernmental decision-making.

Second, the expansion of international law is accompanied by a significant growth of court-like bodies on the international level (Keohane, Moravcsik and Slaughter 2000). International institutionalization seems to follow partly the same logic that unfolded already on the domestic level. This logic implies that the functioning of institutions will depend on the ability to resolve disputes among participants by higher authorities or procedures through which parties can express their views in a dispute. In the past few decades, court-like procedures were less relevant in international environmental politics, but environmental regimes can not be fully uncoupled from the general trend towards the creation of such bodies. The complexity and pitfalls involved with new institutional mechanisms like emission-trading create a certain demand for dispute-settlement. The expansion of these bodies raises the question on whether they can be made accountable to the international community of states or to global civil society.

Third, the growing need for expertise that arises in global governance systems reinforces the long-term development of an empowerment of intergovernmental or transnational elites which are involved in the management of international regimes. Robert A. Dahl's (1994, 33) skeptical prognosis that the third transformation of democracy occurring in connection with the internationalization of governance "will not lead to an extension of the democratic idea beyond the nation-state but to the victory in that domain of de facto guardianship" has primarily been aimed to the level of the nation-state. Global governance systems can also be determined by the evolution of forms of de facto guardianship in the long term if they will fail to provide equal access to participation for different types of non-state actors in world politics.

Regimes as Targets of Change

The conception of international regimes as elements of an international society of states seems no longer appropriate in a world which is characterized by the de-bordering of various spheres of social life. Students of global governance paid broad attention to the impacts of economic globalization or social transformation on the steering capacity of the nation-state in the past decade (Scharpf 1998, Zürn 1998). International institutions were mainly seen as possible remedies to cope with denationalization. But less energy was spent on exploring the impacts of these changes on institutions themselves. Processes like economic globalization, social transformation, or the evolution of new communication technologies indicate the evolution of a world society. Which impacts arise from broader changes in world politics for institution-based order beyond the nation-state? How relevant must the nation-state be considered for the evolution of institution-based order? Martin Shaw (2000, 2–9) illustrates that social scientists developed three broader narratives of change in world politics. Each of these changes impacts on international regimes and produces a new demand to legitimize institutions in world politics. First, postmodern approaches grasp the phenomenon of change under the label of "transformation" considered as a fragmentary and diffuse (instead of unified) process. While these theories refer to the pluralization of cultural, social and political life within and beyond the nation-state, they also overlap with the second strand of literature which understands change as economic, political, or cultural *"globalization"*. David Held and Anthony McGrew (1998, 235–8) take a mediating role between the positions of hyper-globalizers and skeptics, which either predict the demise of the nation-state or declare globalization as a myth. Both authors argue that globalization will not render the nation-state obsolete. But globalization produced a new context for the state and led to the evolution of multiple power centers or overlapping spheres of authority. The perspectives taken by students of transformation and globalization point out that international institutions are confronted with a variety of partly conflicting transnational political attitudes, values and identities expressed by transnational networks, sub-national and local groups or individuals. In this new pluralism all these actors evaluate regime policies, political procedures, outcomes and impacts on the basis of cultural, social, or political understandings. A third approach which conceptualizes change in connection with the "end of the cold war" supports this conclusion. The collapse of the Soviet Empire and the end of East-West-antagonism in 1989/91 illustrate that individual and collective self-determination can gain the upper hand over authoritarianism and heteronomy. But we can also find less encouraging examples in many other countries. They lead us to conclude that we are still far from complete fulfillment of the rights of the individual on the global level.

The formation, evolution and management of international institutions occur in a transnational political environment where global civil society or its different segments exist as responsible "demoi". These "demoi" demand increasingly of these governance systems to provide access to political decision-making procedures and to justify political outcomes. These groups will assess international regimes no longer only in regard to whether regime policies will serve their utilitarian purpose. International governance systems will have to prove whether the procedures underlying the allocation of political values reflect the demand for participation of global civil society. While national democracy rests on a territorially delimited demos, international governance is characterized by functional differentiation. International regimes represent a context within which the traditional understanding of a territorially definable ethnos-demos must be considered inappropriate to serve as a basis for determining a democratic constituency (Wolf 2002). Since it is obviously difficult to develop criteria which define the membership of de-bordered demoi on the transnational level, the question arises whether re-parliamentarization on the domestic level is a possible option that can be chosen to reduce the democratic deficit involved with the internationalization of governance. But it seems doubtful whether re-parliamentarization on the domestic level can fulfill the demands for participation that exist trans-nationally. Sociological developments like transnational community-building and the evolution of new identities suggest the conclusion that demands for democratic participation have to be fulfilled at that level where they arise.

Global governance systems are those arenas in world politics where emerging global civil society politicizes the distribution of social welfare between the developed and the developing world. The issue of distributional justice has gained new relevance at the transnational level. The emergence of global civil society and increasing relevance of cosmopolitan awareness involve, that costs and benefits that arise in connection with the existence of international regimes can no longer be purely considered from the perspective of the national interest. Regime consequences are assessed by states or social groups with respect to the fulfillment of normative claims. These actors assess whether cost-benefit-relations are acceptable to them and whether they represent equitable solutions. The more global governance systems will impact on domestic political value allocation, the more will the issue of distributional justice become relevant as a factor that must be realized in order to secure obedience. The more the policies of international regimes will affect the distribution of political values, the more will affected actors make their willingness to implement, obey, or internalize regime provisions conditional upon fulfillment of demands for distributional justice.

Chapter 3

What is, and How Can We Measure, the Legitimacy of Regimes?

The concept of legitimacy implies that we carefully weigh up which reasons justify obedience to an obligation. This conception of legitimacy goes beyond Max Weber's understanding of the subject in which legitimate rule was reflected in *people's belief* in the three different types of *rational-legal, traditional, or charismatic* legitimacy (Weber 1968, 33–8). The Weberian understanding of the subject has been widely criticized because it separates the belief in legitimacy from the evaluation of political rule against external moral reasons (Beetham 1991, 8–15; Habermas 1992, 541–63; Sternberger 1967). A conception of legitimacy which ignores the weight of morality takes the risk to justify authoritarian political power as legitimate then if the people are convinced of the adequacy of immoral power. If enlightenment and democracy are seen as indispensable achievements of modern civilization, the concept of legitimacy has to consider that external moral reasons also, though not exclusively, justify obedience of the people to political power.

International regimes are problem-driven institutions. They are established for the purpose of responding to urgent political problems in world politics. Transnational problems are predominantly perceived of as a mixture between the problems that occur in the natural world (e.g., the depletion of ecosystems), or in the social world (e.g., social behavior, social traps, diverging constellations of interests). The existence of international regimes can be justified by the utility which they provide to states and their societies. Problem-solving in the natural world can only be achieved by initiating changes in the social world. Efforts which intend to induce changes in the social world must often be carried out although only insufficient knowledge exists about the impacts of these changes on the natural world. It is more likely that an international regime can finally lead to improvements in the natural world, if the regime contributes to improving knowledge about the nature of a problem, if it increases information about options which are available for dealing with a problem, or if it strengthens capacities of member states. International regimes were long predominantly understood as instruments of change in international society. But the rise of global civil society makes it necessary to broaden this view. Grounds which justify obedience to the policies of international institutions can not only be derived from the contribution of these governance systems to problem-solving. There are also normative grounds that become relevant with the evolution of transnational public spheres or with the demand for distributional equity. The distinction between the two dimensions of *input-* and *output-oriented* legitimacy has frequently been used by analysts to establish that obedience to collectively binding decisions can only be justified if these decisions originate from collective will-formation or from the constituency in question ("government by the people") and if the effects will lead to political solutions that are preferable to the outcomes and impacts of alternative forms of cooperation ("government for the people"). Fritz W. Scharpf (1999, 167) concedes that the internationalization of policymaking creates exogenous constraints which limit the range of options for the management of political problems available to national constituencies. But the author is skeptical as to whether an input-oriented perspective can be applied to international institutions like the European Union. He points out that there is a lack of a European "we-identity", that a European political discourse is underdeveloped, and that political associations or media are lacking at the European level. This led him to conclude that one should refrain from evaluating the democratic character of European politics.

The objections put forward against input-oriented analysis of politics in the European Union are valid in a similar way for other international institutions. This skeptical view demands to avoid the transfer of political competence to levels beyond the nation-state and to preserve the autonomy of the national "demos". But it is also doubtful whether such a position can be maintained in the long term. The de-bordering of social and economic life leads to the establishment of global polity-like spheres of authority. These new political spheres make it necessary to develop a new de-nationalized concept of the sovereignty of the people. Therefore, input-oriented legitimacy on the transnational level can no longer be considered irrelevant. Democracy will no longer exclusively be bound to territorial entities like the nation-state. It will be expanded to functional-sectoral systems of policy-making beyond the nation-state. Functional-sectoral forms of international governance can not rely on the cohesiveness of common historic experiences, common culture or language, or a sense of national identity that guarantees the willingness to obey to rules at the national level. Instead, the willingness to obey systems of rule that exist beyond the nation-state can only emerge from the fulfillment of grounds for legitimacy. Such grounds demand that institutions contribute to collective utility or that concerns of all people who are affected by these institutions must be treated equitably and fairly. The distinction between input- and output-oriented legitimacy seems to imply that the two dimensions can be treated independently of each other. But the two dimensions are interrelated. The degree of input-oriented legitimacy which can be found in a transnational polity can also influence fulfillment of output-oriented legitimacy.

In the following, grounds that justify obedience to institutions in world politics will be connected to broader theoretical debates in the study of international institutions. Three grounds for legitimacy are connected to approaches that rely on self-interest or knowledge for explaining the evolution of order in world politics. It will be argued that regimes deserve obedience if they contribute to the reduction of a number of uncertainties in the knowledge-base (1), if they contribute to achieving compliance with regime norms (2), if they have a causal impact on goal-attainment or environmental problem-solving (3). Two other grounds for legitimacy are derived from normative requirements that consider procedural and distributional issues in world politics. Obedience to norms and rules of regimes will then be justified if distributional outcomes and impacts of regime policies reflect normative claims of equity or sustainability (4). Finally, obedience to regime policies is justified if procedures exist which guarantee participation by global civil society (5).

Knowledge, Policy Options, Capacities

One among a number of grounds which justify obedience to international regimes is linked with the contribution of regimes to the reduction of various uncertainties. The utilitarian or instrumental character of international regimes has been emphasized by Robert O. Keohane's (1984) functional theory. This theory considers international regimes as remedies against political market failure. The absence of institutions can prevent states from the achievement of mutually advantageous cooperation. Regimes are understood as contractual relationships between states which create patterns of legal liability that foster the evolution of mutual expectations about the behavior of actors, which alter transaction costs in issue areas, or which provide informational functions and thereby reduce uncertainty in problematic social situations (Keohane 1984, 88–96). The first wave of studies about regime effectiveness was primarily referring to this approach. In particular, it has been argued that international institutions can "help to ameliorate problems of concern, capacity, and the contractual environment" (Levy, Keohane and Haas 1993, 398). This understanding of regime effectiveness focuses on problems which are located in the social world. This understanding is based on the assumption that the creation of institutional mechanisms can contribute to transforming situation-structures which normally undermine cooperation (e.g., lack of trust among members, deliberate cheating, lack of capacities) so that cooperation becomes more likely and stable.[1]

The institutional design is a response to the specific constellation of interests that characterizes the situation-structure. This design can represent a starting point for the development of one among various grounds that justify obedience to regimes. Such a ground for legitimacy can exist if institutional mechanisms and regime policies contribute to reduce or avoid problems in the social world. The functional role ascribed to regimes has first primarily been understood on the basis of a state-centric conceptualization of interests. The observed changes in

---

1   The studies included in an edited volume about the rational design of international institutions highlight that variances in a number of variables that must be ascribed to the social world account for differences in the design of international institutions (Koremenos, Lipson and Snidal 2001a and 2001b). Findings by case studies included in this edited volume support the conclusion that severe distributional problems or strong uncertainty concerning the knowledge about the consequences of possible state action can lead regime members to the creation of flexibility mechanisms (e.g., escape clauses, clauses that permit re-negotiation or further development of existing rules).

world politics lead us to depart from a conception of state interests put forward by traditional rational choice analysts. They assumed that the actors and interests are fixed in analysis and that change occurs mainly as a result of changing constraints in the international environment. Duncan Snidal (2002, 84) has pointed out that "fixed preferences allow for a tight analysis of many issues in an empirically falsifiable way, whereas assumptions of changing preferences lead to slippery and untestable arguments". He also conceded that the interest-based study of international institutions is about to relax the assumption of fixed preferences. The expansion of a traditionally state-centric understanding of interests opens new opportunities for the analysis of global governance systems. It allows adjustment of interest-based approaches to the changes that took place in world politics. The study of the formation and change of state interests converges more to a liberal conception of world politics. This type of analysis considers the de-bordering of politics or the diffusion of transnational actors and it departs from the traditional assumption of the state as an independent actor. It retreats from the assumption that the state will embody the total will of the people or that the societal will is homogeneous and constant (Wolf 1995, 256–9). We can no longer disregard actors like scientists, legal and technical experts, non-state actors or private firms, which influence institutions substantially. These actors contribute to the production of outcomes which can be partly conflicting with those preferred by states.

Before collective action problems can be resolved or policies can be implemented in regimes, actors must be capable of developing their preferences on the basis of knowledge gathered through the programmatic activities of a regime. Normally, states make the evolution of consensus about this knowledge a precondition before they will change their preferences and agree to the creation of a regime. If such consensus has emerged, collective actors and individuals find it easier to change their social behavior towards environmentally sound production and consumption patterns. International regimes can serve as instruments through which members coordinate their research activities, monitoring efforts, or the exchange of data.[2] The scientific and technical collaboration in regimes must be assessed whether it contributes to the development of policy options that can be used for environmental management within and beyond the nation-state. Expertise and capacities needed for the development of policy options are often lacking – particularly in developing countries. If international regimes can only marginally contribute to reducing these uncertainties, they fail to provide a central achievement, for which they have been praised by adherents to knowledge-based approaches.[3] Many countries differ with

---

2   The pooling of originally de-centralized efforts by states for the improvement of the knowledge-base can reduce transaction costs and improve the quality of scientific research or direct the focus of scientific research more to the transnational character of an environmental problem (Andresen, Skodvin, Underdal and Wettestad 2000; Abbott and Snidal 1998, 13–14).

3   For example, Peter M. Haas (1992, 14) diagnosed that decision-makers "do not always recognize that their understanding of complex issues and linkages is limited, and

respect to the availability of financial, technical or scientific capacities required to participate effectively in social practices at the international level. Countries which lack these capacities run the risk that their environmental situation or socio-economic needs will not adequately be considered in the development of scientific consensus about cause-effect-relationships or of policy options. Lack of such capacities for the implementation of policies can lead to the rise of new uncertainties during regime management and constrain effective implementation. Insufficient capacity-building can lead to severe compliance problems that create strong uncertainties on the part of those states which comply. Therefore, capacity-building is necessary for avoiding the evolution of new uncertainties.

Compliance

The twentieth century has been characterized by unprecedented growth of international law that covered nearly all spheres of social life on inter- and transnational levels. The existence of international norms alone hardly tells us anything about the degree to which states and transnational civil society feel themselves bound by them. International politics is lacking central authority which could enforce compliance in such a way as it is practiced domestically. Explanations for compliance were long mainly referring to power- or interest-based approaches. Andrew Hurrell (1993, 53) admitted that the central problem for regime analysts will be to establish that international regimes exercise a compliance pull of their own which will be caused by factors which are at least partially independent of the power and interests responsible for their initial creation. Therefore, the author argued that it will be "necessary to show not only that rules exist and that they are created and obeyed primarily out of self-interest or expediency, but also that they are followed even in cases when a state's self-interest seems to suggest otherwise".[4]

The argument after which the achievement of a sufficient level of compliance can tell not much about the qualitative dimension of social order seems to speak against the inclusion of compliance into the dependent variable. Even if states comply with the provisions of a regime, this must not inevitably lead to effective problem-

---

it often takes a crisis or shock to overcome institutional inertia and habit and spur them to seek help from an epistemic community". The author argued that decision-makers should consult so-called epistemic communities because these actors elucidate the causes and effects or possible solutions to single problems, because they illustrate the nature of complex inter-linkages existing between different issues, or because they can support the development of collective interests and the formulation of policies.

4   Against this background, Ronald B. Mitchell (1994, 30) defined compliance not only as "an actor's behavior that conforms to a treaty's explicit rules". He also distinguished treaty-induced compliance "as behavior that conforms to such rules *because* of the treaty's compliance system" [emphasis by author].

solving, let alone to the production of equitable outcomes. Regimes can fail to solve the environmental problem which they are supposed to manage although the behavior by members meets regime requirements. On the other hand, it is normally inconceivable that a sufficient degree of problem-solving can be achieved without that states will be in compliance with the provisions of a regime. Behavior of social actors is based on reciprocity. Compliance by others can be considered to represent one among a number of relevant grounds justifying obedience to international institutions. Actors which comply expect that their own behavior will cause similar compliance behavior by others.[5] If other actors in world politics comply with rule systems that, by and large, correspond to a trans-nationally shared understanding of effective, equitable, or democratic social order beyond the nation-state, there is a ground for an individual actor that justifies obedience. The state has traditionally been considered as that type of actor which takes on obligations in international treaties. It is responsible for guaranteeing compliance with international norms and for implementing measures which translate them into domestic practices. Which implications have arisen for compliance from the relevance of new actors other than the state in world politics? Which consequences arise for the achievement of compliance from the insight that de-nationalization imposed new constraints on the capability of the state to produce the effects which are demanded by international treaty language? Obviously, the systemic view of international politics which characterized both realist and neo-institutionalist assessments concerning the causal role of institutions for compliance has produced insufficient results. The assumptions of realist theory after which state behavior is determined by relative gains-seeking, by autonomous action, or by disbelief that international norms can be enforced in light of lacking central international authority leads realists to conclude that compliance will mainly depend on the power distribution that prevails on the systemic level. Under the circumstances, the realist assumption that treaty compliance arises from the calculation of costs and benefits leads to the implication "that noncompliance is the premeditated and deliberate violation of a treaty obligation" (Chayes and Handler Chayes 1993, 187).

The consideration of a broader range of factors which are partly located below the systemic level of international politics made institutionalism more aware that noncompliance can also be caused by constraints that can not be attributed to relative gains-seeking of states. These factors contributed to relativizing the realist assumption that noncompliance is primarily the result of deliberate violation. The studies produced by Ronald B. Mitchell (1994) or by Abram Chayes and Antonia Handler Chayes (1993) put forward a managerial approach on compliance. They

---

5   Noncompliance by a single actor can create new uncertainties, undermine trust among members and consequently destabilize social order beyond the nation-state. If the worst comes to the worst, the condition of reciprocity underlying cooperative arrangements can cause a so-called "echo-effect", where other players will do the same in response and "the result would be an unending echo of alternating defections" (Axelrod and Keohane 1986, 245).

argue that noncompliance has long insufficiently been understood to mainly arise from deliberate assessment by regime members concerning their individual pay-off, whereas the causal role of inadvertence for both compliance and noncompliance and of various sub-systemic factors were long neglected as possible explanations for variances in compliance behavior (Chayes and Handler Chayes 1993, 188–97). The managerial thesis about compliance has in turn been attacked by George W. Downs, David M. Rocke and Peter N. Barsoom (1996, 380) who argued that high levels of compliance "result from the fact that most treaties require states to make only modest departures from what they would have done in the absence of an agreement". The authors diagnose that analysts of compliance would be faced with an endogeneity problem because it could be assumed that states will "rarely spend a great deal of time and effort negotiating agreements that will continually be violated" (Downs, Rocke and Barsoom 1996, 398). The assumption of endogeneity implies that limitations exist in regard to the inferences which can be made from compliance data. It is based on the argument that compliance with a treaty is normally explained on the basis of cases where states have *ex ante* decided to comply when they ratified a treaty.

The assumption of an *ex-ante*-decision for compliance is built on a deterministic view of the political process and underestimates the eventuality of economic or political changes which can create new unforeseeable problems in the longer term and make it more difficult for states to comply. The relative distribution of gains – a topic to which realists are very receptive – can not always be completely assessed from the beginning when states will have to decide whether they will become members of an institution. In contrast, the idea of a compliance pull that comes from the existence of a regime arises from the assumption that the course of history is not entirely calculable. Accordingly, institutions are understood to deal with all those factors which impact on regimes and weaken the capacity or willingness of single states to comply with regime provisions (e.g., economic recessions, lacking resources for implementation). While it is doubtful whether the assumption of endogeneity is generally tenable, George W. Downs, David M. Rocke and Peter N. Barsoom (1996) offer an interesting hypothesis according to which increasing depth of cooperation involves a growing demand for enforcement mechanisms. The depth of cooperation is perceived of as the extent to which a treaty requires states to depart from the behavior which they would have taken in the absence of a treaty. This approach would lead us to assume that with growing interference of international and transnational law in social life the likelihood of defection by single actors also increases and therefore requires elements of enforcement. Thus, the authors who suggest an enforcement approach find some evidence for the game theorist prediction "that as regulatory rules tighten, the magnitude of the punishment needed to deter defection would also have to increase" (Downs, Rocke and Barsoom 1996, 391).

Enforcement and management have mainly been treated as competing approaches, but one can also consider them as complementary to one another. This argument put forward by Jonas Tallberg (2002) is built on the observation of a "management-enforcement-ladder" within the European Union that combines cooperative and coercive measures and consists of different elements involving preventive capacity-building and rule clarification, monitoring for the enhancement of the transparency of state behavior and the exposition of rule violation, legal procedures which allow that cases of non-compliance can be brought up and which ensure clarification of existing rules, and the final measure of deterrent sanctions through which compliance can be coerced. The depth-of-cooperation-hypothesis, however, possesses some plausibility not only because defection can be likely in special cases where actors wish to avoid implementation costs. Rather, growing depth of cooperation also supports the relevance of the management approach. The more far-reaching and comprehensive the policies of global governance systems will become, the deeper will implementation on the domestic level affect and interfere in social life. Under the circumstances, domestic implementation of international policies can provoke resistance by sub-national interest groups or be blocked by deeply rooted behavioral practices or domestic institutional properties. The availability of capacities for problem-solving on sub-national levels will gain further relevance (Dai 2007). The existing divide between the developed and the developing world also implies that in many parts of the world noncompliance will (in combination with other factors) be caused by lacking capacities rather than by relative-gains-seeking by states. Such a mix of elements that come from both approaches can also increasingly be found in governance systems that exist outside the European polity. To sum up, compliance by others provides a ground that justifies obedience by oneself. Compliance alone hardly tells us anything about the character or quality of social order in world politics, but without compliance various other grounds for legitimacy can not be realized.

Goal-Attainment and Problem-Solving

Whether institutions are instruments for the achievement of goals in international politics was very much disputed between members of the research community on international politics in the twentieth century. Susan Strange's basic assumption that regimes are epiphenomenal and that they "are only too easily upset when either the balance of bargaining power or the perception of national interest (or both together) change among those states who negotiate them" (Strange 1983, 345) still represents a common point of reference for adherents to a neorealist conception of international politics. John W. Mearsheimer (1995, 7) further stimulated the debate with neo-institutionalists by arguing that institutions "matter only on the margins" and "have minimal influence on state behavior and thus hold little prospect for promoting stability in a post-Cold War world". Neo-institutionalists countered the neorealist challenge by emphasizing that institutions can cause the effective management of trans-boundary problems also independently of the power-structure prevailing in the broader political environment of international politics (Keohane and Martin 1995; Ruggie 1995). The claims for superiority made by single theories did not lead us very far. Lisa L. Martin and Beth A. Simmons (1998, 730) concluded that the debate "has obscured more productive and interesting questions about variation in types and degree of institutional effects".

The question on whether regimes achieve their goals or contribute to environmental problem-solving pertains to the core of a utility-oriented understanding of global policymaking. Global governance systems will be pointless or undesirable and will not deserve obedience if they will not contribute to problem-solving. Grounds for legitimacy that refer to goal-attainment reflect the internal logic of political decision-makers or state officials who long determined the setting of goals in world politics. In contrast, grounds for legitimacy that take problem-solving as a starting point are more based on a logic that arises from an external perspective. In the following, it will become obvious that empirical assessments involve a number of tricky methodological questions that become even more serious in the context of a quantitatively-oriented coding project. These issues will be described extensively in connection with the consideration of the study of problem-solving.

The Goals of Global Governance

Social order is reflected by the achievement of specific goals shared by a collectivity. The term of "governance" implies that various types of institutions will be used as mechanisms to establish social order. James N. Rosenau (1997, 145) perceives governance as spheres of authority "at all levels of human activity – from the household to the demanding public to the international organization – that amount to systems of rule in which goals are pursued through the exercise of control". What are goals in international environmental politics? How can they be identified? What can be gained for the study of regime effectiveness from exploring the degree to which international institutions contribute to the achievement of goals? The goals of a regime can normally be found in the preambles or in those paragraphs of the constitutive documents of regimes which include the basic principles that determine political action. The advantages that are combined with the study of goal-attainment vis-à-vis other widely used standards for the study of regime effectiveness like the notion of the collective optimum (e.g., the Pareto frontier) or the study of compliance were described by Thomas Bernauer (1995, 368). He argues that the standard of goal-attainment is preferable because it "is easier to identify than a collective optimum, which scientists may disagree about and economists may find difficult to determine because costs and benefits are hard to measure".

The study of goal-attainment normally avoids taking an external evaluative perspective which assesses whether the goals formulated by states are desirable or far-reaching enough. While goals provide guidelines for long-term political action in an issue-area, they can be the result of compromise achieved between different groups of states during negotiations which disagree about the degree to which environmental concerns should predominate other (e.g., economic) concerns.

On the other hand, the study of goals allows determining how far other grounds for legitimacy (e.g., economic or normative goals) are reflected in international legal documents. While regimes can address more than one goal, these goals are often causally connected. The framework-convention/protocol-approach that characterized the evolution of international environmental law in the past few decades involved that framework conventions often served to establish institutional mechanisms to promote functional goals (e.g., improvement of international cooperation in the issue area, increasing scientific collaboration and information exchange), while more attention has been given to the achievement of environmental goals (e.g., protection of the marine environment of regional seas) by those legal documents (e.g., protocols or annexes) that followed in the aftermath. Therefore, it seems plausible to distinguish between functional and environmental goals. Functional goals apply to those specific regime functions which are necessary to achieve a sufficient degree of certainty regarding the causes and consequences of a problem (e.g., scientific research or monitoring), which contribute to reduce transaction costs or enable single states to participate in social practices (e.g., capacity-building) or which generally strengthen cooperation in the issue-area. Environmental goals specify which state of the world should be accomplished in the long term. In addition, other goals which lie beyond the categorization of functional or environmental goal-orientation have frequently been established in regimes. For example, they can call for the fulfillment of moral claims like distributional justice or increased participation of transnational civil society.

Problem-solving: Relative Improvements or the Collective Optimum?

This dimension of regime effectiveness primarily refers to problems in the natural world that have been caused by human behavior. The depletion of stratospheric ozone or the pollution of regional seas represent issues for which environmental scientists can approximately determine how far the measurable state of the environment is away from ecological equilibrium. Trans-boundary emissions of air pollutants, the export of toxic waste, trade in endangered species, or the amount of living resources that can be found in single ecosystems can be measured or calculated more or less reliably by the use of standardized methodologies and technologies that became available for the exploration of environmental problems. The study of problem-solving effectiveness centers around a number of basic issues (Helm and Sprinz 1999; Mitchell 2001; Underdal 2002a; Young 2001): Would the state of the world be different in the absence of a regime? How must observed levels of change achieved by a regime be assessed against the non-regime-counterfactual? To what degree are regime and non-regime factors accountable for observed levels of regime effectiveness? Is it possible to assess these levels of changes achieved by a regime against the possible optimum?

Two different approaches concerning the measurement of regime effectiveness are reflected in these questions. First, the approach that explores the *relative effectiveness* of regimes involves comparison of the achieved level of problem-solving with the hypothetical level that would have occurred in a non-regime situation. The second option involves to assess the actual level of problem-solving against a hypothetical solution that could have been achieved in *ideal circumstances*. This option takes the Pareto frontier as a maximum standard that is reached "when no further increase in benefits to one party can be obtained without leaving one or more prospective partners worse off" (Underdal 2002a, 9). The collective optimum represents the more interesting point of reference for students of regime effectiveness. It confronts global policymaking with more ambitious demands than the other approach, because the possible optimum sets a very high standard against which actual achievements must be assessed. The collective optimum suggests that grounds for legitimacy will be closely connected with a conception of an ideal solution that will serve as a benchmark with which actual achievements will be compared. In contrast, the non-regime counterfactual takes on the character of a worst-case-option against which actual achievements can sometimes appear in a good light. Accordingly, this approach is based on less demanding claims that can be used to justify obedience.

Which of the two approaches serves the purpose of a comprehensive coding project? First of all, both approaches do not fully correspond to political reality. The optimum approach runs the risk of directing too ambitious demands to global policymaking that may be justified from an environmentalist perspective but come into conflict with needs for social and economic development or the demand for economic efficiency. While the collective optimum is often unknown at the beginning of political management, later periods of collective decision-making can also still be characterized by a significant degree of uncertainty in this respect. A conception of grounds for legitimacy that is based on social reason requires that the standard against which problem-solving will be measured is a reasonable option. Obviously, the availability of an unquestionable collective optimum is not always guaranteed. On the other hand, relative improvements can be so insufficient that no significant differences can be observed compared to previous environmental conditions. Accordingly, grounds that justify obedience with reference to relative improvements will correspond to social reason only if perceptible improvements will be made by international institutions. Before we can answer the question about the suitability of the two approaches for a comprehensive coding project, a critical examination shall be made which balances the constraints of the two approaches for the quantitative study of regime effectiveness.

While the relative improvements approach and the collective optimum have first been considered as two alternative options for studying regime effectiveness, Carsten Helm and Detlef Sprinz (1999) made an attempt to merge the two approaches. They developed an effectiveness score on the basis of three variables: the no-regime counterfactual (NR), actual performance (AP), and the collective optimum (CO). Regime effectiveness is perceived as the ratio between the numerator AP *minus* NR and the denominator CO *minus* NR. The two authors tried to demonstrate the usefulness of their approach by using the issue area of trans-boundary air pollution in Europe as an example. In determining the upper and lower bounds of hypothetical emission reductions that become part of the above-described ratio, the authors rely on a number of data sources that are available from international monitoring facilities or sophisticated simulation models. Further on, their findings are based on other research techniques like expert interviews. The authors calculated possible reductions of emissions of sulfur dioxide and nitrogen oxide that occurred in more than twenty European countries for the two hypothetical baselines of a no-regime counterfactual and the collective optimum and confronted these figures with actual reductions. These reductions have been achieved in the 1980s and early 1990s on the basis of the 1985 Helsinki Sulfur-Protocol and the 1988 Sofia NO_x-Protocol which both emerged in the context of the 1979 LRTAP-Convention. The findings produced by this study suggest the conclusion that the "aggregated regime effectiveness scores are substantively larger than zero in both pollutant domains but fall short of their theoretical maximum" (Helm and Sprinz 1999, 16).

On the first view, this approach seems to offer an elegant solution for measuring regime effectiveness. To some extent, the usefulness of this approach depends on the character of the problem that will be considered by analysts. In general, emissions are more suited for the modeling of alternative scenarios within temporally limited periods of time, whereas other environmental problems like desertification or biodiversity must be assumed to be less qualified in this respect, because achievements can by far less be measured in calculable units. However, the approach can also be used to illustrate the limitations that arise for the study of regime effectiveness in the context of a comprehensive coding project. The above-described approach would demand too much of coding case study experts, because it involves to deal with two hypothetical baselines simultaneously. It includes the danger of speculative judgments or provokes refusal of coding case study experts to determine the degree of problem-solving effectiveness in the issue-area because they may feel unable to determine the collective optimum reliably. The production of reliable data by a coding project which builds on the expertise of many different case study experts will depend on the development of inter-subjective understandings among coders regarding the logical steps that must be taken or the variables that must be considered for arriving at conclusions concerning the character of a collective optimum. Since we still lack a conceptual framework for determining the collective optimum that could be used across the boundaries of single issue areas, the collective optimum had to be ruled out as an option for research-pragmatic and methodological reasons.

Which conclusion can be drawn for the empirical study of relative improvements in the state of the world? In order to make codings about problem-solving effectiveness comprehensible, empirical observations on changes in the state of the world must be separated from causal judgments about the relevance which can be ascribed to regimes for these changes. Thus, a first measurement must be made which indicates whether the state of the world has changed during a specific time period. This measurement assesses whether the problems managed by a regime improved or worsened or were determined by continued existence of the status quo. Only in a second step, a causal judgment can be made which considers whether changes in the state of the world have been caused by regime factors or by external factors and which possible developments would have occurred in the absence of a regime. The methodological difficulties that arise from the use of the no-regime counterfactual as a basis against which actual achievements can be assessed should not be downplayed. With respect to the usefulness of counterfactual reasoning, James D. Fearon (1996, 66) remarked "that for typical social science problems we will only be able to judge the plausibility of counterfactual arguments for highly 'local' situations [and that] we will be able to assess plausibility only where the counterfactuals invoke causal mechanisms and regularities that are well understood and that are considered at a spatial and temporal range small enough that multiple mechanisms do not interact, yielding chaos". Therefore, the longer the time horizon that characterizes the evolution of a regime or the more comprehensive the number of variables which must be taken into account by counterfactual analysis, the more difficult it will be to gain reliable findings. What speaks in favor of the no-regime counterfactual from a methodological point of view is the existence of an identifiable historical moment where the non-regime-situation has changed into a regime situation. This event can be taken as a starting point for developing a scenario that further rests on the hypothetical assumption of a non-regime-situation and models an alternative course of history in the issue-area. While both approaches raise a number of serious methodological problems, the no-regime counterfactual is only the less problematic alternative that exists in this respect. To sum up, goal-attainment and problem-solving are two different approaches that can be used for measuring similar, though not identical, dimensions of problem-solving effectiveness. The methodological pitfalls involved with the study of a collective optimum lead us to conclude that an assessment of the relative improvement or worsening of a problem seems to be the approach which is more reliable.

Distributional Consequences and Justice

Rationalism considers pay-offs which states can expect from institutions as a central factor for the evolution of preferences, but normative considerations which assess whether these impacts are equitable have long been ignored by this strand of literature. The equitable character of distributional outcomes that possibly arise from governance systems is often considered as a precondition by states and societal actors to follow and implement international rules. Political theorists disagree concerning the meaning of the concept of justice outside the realm of the nation-state. While John Rawls' theory of justice applies to national democracies, it has been used by some students as a starting point for exploring how far justice can be realized also in international politics. The Rawlsian theory of justice is based on the hypothetical assumption of an original position where individuals act under the veil of ignorance and lack knowledge about their own status, sex, religion, class, or talents or other factors which could bias the production of fair outcomes. Against this background, Rawls arrives at two principles which characterize his theory of justice with the first principle having priority over the second. The first principle involves that each person has an "equal right to the most extensive total system of liberties compatible with a system of liberty for all". The second principle determines that inequalities that occur in social or economic respect shall be "arranged so that they are both: (a) to the greatest benefit of the least advantaged, consistent with the just savings principle, and (b) attached to offices and positions open to all under conditions of fair equality of opportunity" (Rawls 1971, 301).

By making explicit reference to Rawls' theory Charles R. Beitz (1979, 128) argued more than two decades ago, that "international relations is coming more and more to resemble domestic society in several respects relevant to the justification of principles of (domestic) social justice". The author suggests that these principles can no longer be limited to the domestic level but ought to be applied globally. The efforts made by Charles R. Beitz or Thomas Pogge to translate the Rawlsian approach into a cosmopolitan concept were assessed by Molly Cochran (1999, 46–51) who illustrates a number of difficulties that arise in this context. *First* of all, such efforts have to clarify the concept of the person. While cosmopolitanism is based on the idea of liberal individualism which emanates from the assumption of the pre-sociality of the individual, it underestimates that individual identity is also contingent on its social embeddedness. Against this background, Cochran (1999, 46) reproaches cosmopolitans for not taking into account that "the later Rawls accommodates an understanding of the person as being socially chosen as well". *Second*, the cosmopolitan approach implies that the individual rather than the international state system will be considered as a unit for which transnational justice must be realized.[6] Admittedly, the process of transnational societization suggests that demands concerning fulfillment of moral claims on the level beyond the nation-state can no longer be limited to the state-system alone. On the other hand, states are still actors which possess a significant amount of competences for the evolution of social order in world politics. Therefore, cosmopolitanism must clarify whether the evolution of government-like structures beyond the nation-state is understood to replace or complement the nation-state and how far the state or actors other than the individual will be subject to moral demands in world society. The *third* difficulty arises from the efforts made by cosmopolitans to expand the Rawlsian conception to the universal level. For example, Cecilia Albin (2001, 12) rejects the Rawlsian assumption of the veil of ignorance as a possible starting point for the evolution of a conception of justice that can be applied to the level beyond

---

6   From the perspective of developing countries, Yash Tandon (2001, 216) criticized Rawls' refusal to accept that communities and nations can have similar rights than individuals. While accepting that part of liberal ethics which demands the allocation of basic human rights to individuals, the author argues against Rawls methodological individualism because it "denies that societies are more than the sum of individuals".

the nation-state by arguing that it "cannot be operationalised in international encounters which fail to meet the Rawlsian criteria of a fair selection situation".

In light of the difficulties that are involved with the transference of Rawls' normative theory of justice from the domestic to the transnational level, it seems plausible to develop our criteria for distributional equity on the basis of a more inductive understanding that is normally applied by states or collective actors for assessing the distributional consequences of international regimes. Possible points of reference against which the equitable distribution of values can be assessed are often included in regimes. States develop legal principles and norms that demand equitable solutions or set up issue-specific guidelines for the implementation of policies addressing those basic distributional aspects that characterize the issue area. Further on, distributional justice deserves to be treated equally important than the issue of democratic participation. The strong discrepancies that exist with respect to the socio-economic living conditions in world society hardly justify to give procedural fairness priority over distributional justice.[7]

How practicable is the concept of sustainable development as a starting point for assessing the equitable character of international regimes? How far can we follow the cosmopolitan demand that distributive justice should primarily be assessed on the level of the individual? Thomas M. Franck (1995, 58–79) illustrates that equity has become an important principle in international politics in the past century that influences, even though not fully determines, the allocation of resources between states. From a legal-empirical perspective, the author distinguishes three different qualitative dimensions of equity practiced during this period. First, the rather weak form of *corrective equity* has been implemented within single trading regimes like GATT where industrialized countries agreed to establish the Generalized System of Preferences (GSP) in order to weaken, though not fully abolish, the negative impacts which arose for developing countries from the "Most Favored Nations" clause. Second, the more *broadly conceived model of equity* is no longer characterized by pure correction of existing rules, but it perceives equity itself to become an important rule of law as it could be observed in the context of UNCLOS III where the principle of equity finally prevailed in negotiations about maritime delimitation. A third category involves all those resources that have been declared as a *common heritage of mankind* and where equity, like in the case of Antarctica, is embodied by principles that declare non-ownership of the heritage, shared management and benefits, conservation, or peaceful use of the common resource. The differences that exist between these three qualitative dimensions make it obvious that equity has been unevenly realized by international society in various governance systems.

The concept of sustainable development has been introduced by international environmental lawyers to serve as a benchmark for assessing also the equitable

---

7   The adaptation to and mitigation of climate change involve serious questions about justice that divide developed and developing countries (see Adger, Paavola, Huq and Mace 2006).

character of environmental institutions. In particular, sustainable development involves demands for fulfillment of both intra-generational as well as inter-generational equity on the transnational level. But our methodological tools for determining the inter-generational effects of international institutions are still underdeveloped, nor is a sufficient number of economic studies available which calculate the equitable distribution of the outcomes and impacts produced by international institutions from an intra-generational perspective. The cosmopolitan claim after which distributional effects of international regimes must be assessed primarily on the level of the individual can hardly be fulfilled at the moment. Rather, we will have to content ourselves with empirical findings that primarily focus on the distributional consequences occurring on the inter-state level. Consequently, we will have to rely on more descriptive categories on the basis of which it will be possible to develop a general assessment concerning the fulfillment of equity by global governance systems. A quantitatively-oriented approach must take into account that the benchmark for determining the equitable character of outcomes will to a large extent be contingent on issue-area-specific considerations. For example, distributive issues in negotiations over trans-boundary air pollution in Europe have been described by Cecilia Albin (2001, 57) to involve topics concerning "who should undertake emission cuts (given inequalities in past, current and future projected emission levels) and at whose cost (given differences in resources, responsibility for the problem, and in gains to be had from regulations), and by how much emissions should be reduced and by when (given at once the high costs of abatement and the devastating effects of pollution on some countries)".

While it becomes obvious that the possible equitable outcome will largely depend on issue-specific factors, a descriptive approach must rely on a number of basic criteria that can be applied for assessing distributional consequences in environmental issue-areas. First of all, indications with respect to whether distributional justice has been realized as a topic by regime members can be found in the *principles and norms* of an institution. While one has to admit that legal principles like "equitable use of natural resources", "common heritage of mankind" or "intergenerational equity" alone will not guarantee the production of equitable outcomes, their existence indicates that equity has been considered as a relevant topic by regime members. They provide a starting point for exploring how far these legal provisions were further operationalized by regime members or led to policies that were designed with the intention to achieve more equitable outcomes. Second, the identification of distributional consequences considers the *costs and benefits* that occur in the issue area. Under the circumstances, it has to be considered whether the distribution of costs reflects differentiated responsibilities of states for environmental problems and whether it takes different levels of socio-economic development of regime members into account. Third, *assessments of the distribution of costs and benefits* will be carried out for the inter-state level and for the level of important social groups on sub-national or transnational levels. To sum up, various reasons were given which illustrate why it is nearly impossible at the moment to use normative theories of justice as a starting point for developing indicators for the measurement of distributional consequences. Under the circumstances, a procedure will be chosen that can hardly do justice to the ambitious theoretical claims of various normative theories of justice. Nevertheless, exploring the costs and benefits occurring in regimes or a study of the legal principles and norms can improve our understanding with respect to whether normative requirements are reflected in regime consequences.

Transnational Democracy

International law has increasingly been used to internationally codify the claim of individuals for fulfillment of fundamental rights on the domestic level. Thomas M. Franck (1992) diagnosed the emergence of a "democratic entitlement" where only democracy will validate governance on the level of the nation-state. In this context, international institutions are used to monitor and assess how far normative entitlements of the individual to self-determination, to free expression or to participation in a national electoral process will be fulfilled. While it has been argued that our world is "witnessing a sea change in international law, as a result of which the legitimacy of each government someday will be measured definitively by international rules and processes" (Franck 1992, 50), many years will go by until this entitlement will lead to the desired effects also in those parts of the world which lie beyond the western hemisphere. The evolution of trans-governmental networks in the context of global governance confronted the world of liberal states also with new challenges for the democratic process. The moral claims raised by international law towards fulfillment of democracy on the domestic level apply to international governance systems themselves. How far can transnational democracy fulfill the participatory claim of the individual or of collective actors in world politics against the background of contextual factors that characterize global governance systems?

A democratic concept of international law must consider the principle of congruence after which those who will be affected by international norms should be able to participate in decisions that are connected with the creation or interpretation of norms. In modern liberal democracies, the principle of congruence has been realized by the establishment of democratic rights and procedures which guarantee that individuals and social groups can influence political processes. National boundaries delimit the geographical scope of these rights. It is obvious though that national democracy rests on conditions that are either less developed or completely lacking on the transnational level. Obedience to international law can only be justified if this law emerges from, or will be applied and interpreted on the basis of, procedures that are open to participation of emerging global civil society. From a moral-philosophical perspective Otfried Höffe (1997) considered the growth of international law as an element of a more comprehensive process of global constitutionalization. This perspective emanates from the assumption that the breakthrough of social reason which found expression in the rule of law on the domestic level will be followed by a similar development on the level beyond the nation-state. From this perspective, the strengthening of global reason will be fulfilled by the global rule of law and be materialized in the creation of a "minimal world state", even though not in a world government that would hold all-embracing authority, to which all other units in world politics will have to submit. For the time being, we will have to content ourselves with a less utopian conception which avoids evaluating international regimes against the background of a wishful notion of global democracy embodied by whatever form of global statehood. International regimes are often stand-alone institutions which are not subject to decision-making power of an international organization. While cosmopolitans suggest that international organizations should be parliamentarized (Archibugi and Held 1995; Falk and Strauss 2001; Held 1995), this idea misses the truth of many regimes that exist without closer linking to international governmental organizations. Conceptions of global democracy that suggest the parliamentarization of existing international organizations in parallel with the consolidation of government-like structures beyond the level of the nation-state disregard the currently de-centralized character of global governance (Wolf 2002, 36–8).

The fragmentation of international environmental regimes involves that no organizational unit currently exists on the international level that could bring together the broad number of regimes and which could be used to parliamentarize global environmental governance. This implies that a conception of transnational democracy that relies primarily on representative democracy and on the parliamentarization of international governance systems is currently not applicable to environmental regimes. The scope of many regimes is too narrowly defined that the creation of sectoral parliaments could appear as a practical solution. Instead, international regimes can be understood as crystallization points within and around which the international society of states and global civil society are engaged in continuous discourse about the appropriateness of regime policies or the procedures by which these policies have been agreed upon (Bohman 1999). International regimes can be perceived as transnational public spheres where various types of actors deliberate about the rightness and appropriateness of policies or procedures used for collective decision-making. Seyla Benhabib (1995 and 1997) illustrates that a modern understanding of public spheres can be distinguished from Hannah Arendt's traditional conception in two respects: *i)* it takes into account that modern political discourse is no longer primarily based on face-to-face-exchanges but on *impersonal* communication; *ii)* it is *de-bordered* in the sense that it goes beyond the territorial boundaries established by nation-states. In a de-bordering world, the "anonymous public conversation" that characterizes the modern public sphere takes place in a de-nationalized context where communication between stakeholders performs across national boundaries, cultures, and languages. Accordingly, a transnational public sphere can be perceived as a socially constructed space of communication that involves a more or less virtual character and eludes territorial delimitation.

The transmission of discourse from functional-sectoral governance systems to the broader transnational public still faces a number of constraints. The subjects which are discussed in governance systems are often very specialized. The complexity of cause-effect relationships that characterizes some environmental issues makes it often difficult to communicate discourse to a broader audience. The functioning of public spheres is also dependent on the existence of transnational media which transfer discourse from the international regime to the broader global public and vice versa. The evolution of such "relay-stations" (Neidhardt 1994, 11) is a necessary condition for the functioning of public spheres. There are various channels through which a transnational public can communicate and which facilitate that these issues will be perceived. Transnational communication in networks of scientific and technical specialists can rely on the resources that are provided by research institutes, universities, national and international bureaucracies, activist nonprofit interest groups or private companies.

It must be conceded that on the level of international regimes the focus will primarily be directed on the granting of rights for collective associations. The contextual factors that characterize international regimes make it difficult to grant direct participation rights to individuals (e.g., voting rights) in these institutions. It seems plausible to focus at the moment on the development of participation rights for collective actors. For the future, it would be desirable if direct access of both individuals and non-state actors to international courts or to dispute settlement mechanisms in global governance systems could be improved. The main option currently available to individuals who are affected by the impacts of international regimes involves to engage in the activities of non-state actors. Against this background, non-state actors will have to improve the accountability to their members and must be open to the participatory claims from outsiders.

Grounds for legitimacy result from the democratic character of the procedures offered by global governance systems to various types of non-state actors. Which normative principles exist that can be used for determining the democratic character of a regime polity? The formal and informal rules that govern participation of non-state actors are attributes of the regime polity which makes an offer to global civil society to participate. The distinction between formal and informal rules is useful because it allows to determine whether participation rights have been codified by regime members and have taken on constitutional character. The policies which flow from international regimes will be justified if transnational discourse will realize a number of normative conditions that are fundamental to public spheres in the modern era. The concept of the public sphere that has been developed for the domestic level is built on normative requirements that can be used for determining the democratic quality of political processes (Peters 1994, 46–7). The normative core of this concept applies to domestic and transnational publics. The first criteria of *equality and reciprocity* demands that public communication shall be open to each actor that is willing and capable to contribute to the exchange of speech acts. Political communication can not be understood as a one-sided or unequal matter but involves that senders and receivers will be prepared to take both roles. Second, the criteria of *openness* involves that relevant issues will become subject to public debate. The more the regime polity will guarantee that relevant matters in the issue-area will be considered or become subject to public discussion, the more can it be considered as a legitimate arena. Third, the criteria of *discursiveness* imply that arguing characterizes political communication and that participants in a discourse accept and respect each other. Since political communication should be characterized by a joint search for the better argument, persuasion instead of bargaining is considered as the dominant mode that should determine global discourse.

These normative demands allow the determination of the types of rights which must be allocated by a regime polity. Further on, these normative requirements direct our view towards the hierarchical relationships that determine the relationship of societal actors in the context of international regimes. Regimes can be determined by the empowerment of an elite of scientific, technical, economic, or legal experts. The existence of elites alone certainly does not speak against the democratic character of global governance. The empowerment of elites would be problematic if the selection of these elites would be based on coincidence but not be governed by commonly shared criteria like expertise or if their mandate would not be limited to a clearly defined term. The demand for effective functioning of single regime bodies justifies that the regime polity will distinguish participation rights on the basis of criteria that consider the different functions of non-state actors. Participation rules should prevent the self-empowerment of governmental or transnational elites, but they also have to guarantee that regime functions can be carried out in a sensible and effective way. While the status of an observer would put non-state actors only into the position of a receiver, other rights will be required that non-state actors can communicate societal interests to, or participate in political discourse with, decision-makers in negotiation bodies. In this context, access of non-state actors to the documents of decision-making bodies is a right through which it will be guaranteed that transnational civil society will be granted the same level of information available to governments. Further on, the right to speak or the right to submit proposals to state delegates in meetings of decision-making bodies which are given to non-state actors are both indispensable in order to give voice to the arguments of those parts of transnational civil society that are affected by regime policies. In addition, consideration of the demands of global civil society can only be sufficiently achieved if these actors are given a right to suggest own topics for the agenda of decision-making bodies. To sum up, transnational democracy can not rely on elements of representative democracy that determine the democratic process on the domestic level. Regimes must be understood as transnational public spheres where various types of actors engage in political discourse. These systems of political rule justify obedience if the regime polity offers procedures for political participation of emerging global civil society that fulfill a number of normative requirements like equal access, openness, or discursiveness.

Chapter 4
Non-State Actors and the Legitimacy of International Regimes

The assumption that non-state actors can improve the legitimacy of global governance found its way into the rhetoric of global policymakers. The General Assembly of the United Nations resolved in its Millennium Declaration to "give greater opportunities to the private sector, non-governmental organizations and civil society, in general, to contribute to the realization of the Organization's goals and programmes" (United Nations A/55/L.2). The Declaration on Sustainable Development which was adopted by the Rio+10 summit in Johannesburg in September 2002 acknowledges that sustainable development requires "broad-based participation in policy formulation, decision-making and implementation at all levels". The declaration explicitly emphasizes that "to achieve our goals of sustainable development, we need more effective, democratic and accountable international and multilateral institutions" (United Nations A/CONF.199/L.6/Rev.2). This causal connection is reflected in our prime hypothesis that the legitimacy of these governance systems can only be improved if non-state actors can bring their potential to bear. Accordingly, analysis must consider the different functions which non-state actors can have during various stages of the political process.

Global governance leads to legitimacy problems on different levels of world politics. The accountability of transnational elites to the broader transnational public is a crucial issue that must be resolved by various types of non-state actors. The more these actors are included in the decision-making or implementation of international regimes, the more will the question about their democratic mandate become relevant (Beisheim and Zürn 1999, 314). In this study, the focus of analysis will be directed on the legitimacy of international regimes but not on the legitimacy of non-state actors. Both topics are certainly relevant for the evolution of social order in world politics. But each of these topics represents an independent research question. This study will focus on exploring how non-state actors can influence the legitimacy of international regimes. Less attention will be paid to the legitimacy problems that arise on the level of non-state actors.

The analyst is running the risk of glorifying the role of non-state actors, if he/she adopts the prime hypothesis uncritically. We must admit that the claim made by many non-state actors or by political theorists that involvement of these actors in international regimes will raise the quality or democratic character of global governance can be one-sided. Such claims can disregard the problems that emerge in connection with increased participation of non-state actors. Kal

Raustiala (1997, 720) pointed out that "increased participation may also impair regime effectiveness, create policy gridlock (...), and lead to poor outcomes from an environmental perspective". States still play a very important role in global environmental governance. But transnational societization forced states to partly change decision-making procedures or policy styles which they apply during the creation or management of international institutions. Subsequent empirical chapters will demonstrate whether the prime hypothesis is correct or whether it is expression of wishful thinking that does not correspond to the reality.

In the following, two broader steps will be made. These steps contribute to structure the empirical exploration of the impact of non-state actors on international environmental regimes in subsequent chapters of this book. In a *first* step, the understanding of non-state actors will be clarified. The types of non-state actors that were considered for data collection will be described. Analysis will focus on the differences that exist regarding the contributions, strategies and political behavior of non-state actors in global governance systems. In a *second* step, two theoretical approaches will be used for explaining the impact of institutional mechanisms and of non-state actors on the fulfillment of various grounds for legitimacy. An institution-based approach explains the fulfillment or non-fulfillment of grounds for legitimacy as a result of the institutional design. The other approach explains changes in our dependent variable as a result of the activities of non-state actors.

Regimes and the Pluralism of Global Civil Society

Regimes are those arenas where states negotiate on issues which affect the collective interest of emerging global civil society or the utility of transnational economic actors. Discourse about the rightness of arguments and about the appropriateness of policies takes place in transnational public spheres that emerged around regimes. Regimes are those arenas of political management where non-state actors can influence the development or implementation of transnational policies. More than a century ago, non-state actors began to lobby for the creation of international institutions and began to participate in the management of transnational policies. A team of authors studied the evolution of environmental NGOs from a long-term sociological perspective between 1875 and 1990 (Frank, Hironaka, Meyer, Schofer and Tuma 1999, 84–90). The empirical data provided by this group suggest that a few of these actors emerged already in the late nineteenth century. But rapid growth of these organizations occurred only in the second half of the twentieth century. The 1972 United Nations Conference on the Human Environment and the 1992 United Nations Conference on Environment and Development were both accompanied by increases in the number of environmental non-state actors.[1]

---

1   The evolution of non-state actors has affected nearly all spheres of social life and policymaking on national and transnational levels. Although this process has caused positive effects in many respects, it was accompanied in some areas also with negative

The concept of global civil society transfers the traditional understanding of civil society developed for the modern state to the transnational level.[2] Paul Wapner (1997, 66) defines global civil society as "the domain that exists above the individual and below the state but also across state boundaries, where people voluntarily organize themselves to pursue various aims". Wapner's concept of global civil society is based on the assumption that different types of actors with diverging interests exist alongside one another in world politics. It focuses on the existence of various types of associations, whereas the role of individuals is disregarded. It takes into account that noncommercial non-state actors influence policymaking in international institutions. It also considers growing expansion of the global market and the role of transnational firms as political actors in global public spheres.

Some students argue that their own approach is incompatible with the concept of global civil society for conceptual or ideological reasons. For example, Margaret E. Keck and Kathryn Sikkink (1998, 33) pointed out that the concept about advocacy networks cannot be subsumed under notions of transnational social movements or global civil society. They argue that theorists who predict that a global civil society is emerging from economic globalization or rapidly expanding communication and transportation underestimate the issues of agency or political opportunity. Instead, the authors prefer a conception of transnational civil society that is perceived of "as an arena of struggle, (or as) a fragmented and contested area". But adherents to the concept of global civil society are also aware that relations between various types of transnational actors are not only characterized by harmony. Differences exist among the many approaches concerning the range of non-state actors which they include or the political views which they reflect. The concept on global civil society focuses on a broad range of actors, but it ignores the role of individuals or international organizations.

The evolution of a societal sphere on the domestic level has been described by political theorists as a process that involves separation of society from the state with common identities shaping new forms of societal community-building. The concept of global civil society transfers the traditional understanding of civil society developed for the modern state to the transnational level. But this is not unproblematic. The form of state authority which took shape on the domestic level is nonexistent on the level beyond the state. The international state-system is made up of independent units which lack the authority to enforce the monopoly of violence on the international level. Global civil society is not fully independent from states.

---

consequences. For example, the evolution of transnational criminal organizations produced new dangers for individual security or social welfare beyond the nation-state (Kohl 2002, 331–2). Some of these criminal actors established transnational black-markets for the trading of products such as narcotic drugs, various sorts of weapons, or organize modern slave trade across national boundaries.

2   For a discussion of conceptual issues combined with the study of global civil society see also Alkoby (2006).

It interacts with states and it is permeated by legal norms or governmental or semi-governmental organizations which emerged on national and international levels. Further on, the world of non-state actors is determined by the divide between commercial and noncommercial actors. The discussion about the antagonism between transnational nonprofit interest groups and commercial actors is more or less a repeat of the debate carried out for the domestic level, where it was disputed whether the concept of civil society should include or exclude the economy as an actor (Cohen and Arato 1992, 74–5).

Economic and nonprofit actors follow different logics that shape their behavior during the formation or evolution of regimes. Economic actors primarily intend to maximize their own utility. Nonprofit-oriented interest-groups normally aim to promote public goals on the transnational level. But they also intend to increase own material resources which they need for political campaigns. Many theoretical approaches distinguish between the spheres of nonprofit-interest groups and transnational firms. Many nonprofit activist interest groups claimed that the goals which they follow are morally more justified than those of commercial actors. The global economy has begun to harmonize economic profit-seeking with ethical values. But this effort is still in its infancy. Economic and environmentally-oriented non-state actors are not homogeneous groups. Coalition-building across the borders of both camps can be observed quite frequently (Breitmeier and Rittberger 2000, 153).

What exactly do we mean by non-state actors? Priority will be given to the term "non-state actors" vis-á-vis the term "NGOs". The term "NGOs" refers to nonprofit actors, but it fails to consider the sphere of the global economy. Collaboration and coalition-building between economic and environmental non-state actors can be observed just as frequently than conflicts that occur within the realms of these two types of actors. The term "NGOs" also disregards the role which international organizations or programs can play in connection with other non-state actors for the evolution of social order in world politics. In a strict sense IGOs can not be considered as transnational societal actors. But the boundaries between various transnational societal actors are blurring. Transnational social movements or advocacy networks occasionally form alliances with multinational companies or international organizations in order to pressurize the international community of states. It is therefore hardly plausible to neglect the influence of IGOs, because these actors can influence the evolution of state preferences. In the following, different types of non-state actors will be described. These types of actors will be considered by empirical analysis in subsequent chapters.

Nonprofit Activist Groups

Nonprofit activist non-state actors can take different roles as advocacy or service organizations (Gordenker and Weiss 1996). Advocacy organizations initiate public discourse, criticize the outcomes of global governance systems, or demand implementation of specific policy solutions. They play a more political role than service-oriented actors, but the latter of both types of actors is not apolitical. The influence of nonprofit activist groups is not limited to the early stage of agenda-setting. Transnational environmental activist groups follow and try to influence international negotiations (Betsill and Correll 2007). They transmit information about the negotiation process to the transnational public or exercise public pressure on negotiators. These actors raise awareness for problems that occur in connection with the ratification or implementation of, and compliance with, regime policies. These non-state actors use a broad spectrum of political strategies. Fundamentalist actors do not exclude the use of civil disobedience. Pragmatic environmental actors use less spectacular forms of political action.

Nonprofit activist groups exist more or less independently of states. They intend to induce long-term changes in transnational political values. These changes in values can contribute that environmental issues will be more adequately considered on domestic and transnational political agendas. Their activities are not limited to the level of international environmental regimes, but they can also affect national or local politics. National and transnational societies are just as relevant as addressees of their work than national governments. The protest raised on the occasion of the Third Ministerial Conference of the World Trade Organization (WTO) in Seattle in late 1999 can be understood "as a form of global political contestation" (Kaldor 2000, 106) where emerging global civil society expressed its dissatisfaction with the outcomes of, or the decision-making procedures used by, international institutions. The protests that occur repeatedly on the occasion of international governmental meetings indicate that a global transnational opposition emerged which disagrees with the policies of global economic or environmental institutions. Some students argue that these protests reveal the evolution of global social movements which consist of "groups of people around the world working on the transworld plane pursuing far reaching social change" (O'Brian, Goetz, Scholte and Williams 2000, 13). Global social movements share common transnational identities and are not members of the elites in their societies, but they are perceived of as anti-systemic.[3]

Environmental activists normally intensify their pressure on states with the aim of making international environmental regimes more effective, equitable or democratic, but they are less interested in abolishing or replacing institutions by whatever type of alternative governance system.[4] Private ownership of the means

---

3   Sidney Tarrow (1998, 184) defines transnational social movements as "sustained contentious interactions with opponents – national or non-national – by connected networks of challengers organized across national boundaries". The author emphasizes that the challengers themselves are "both rooted in domestic social networks and connected to one another more than episodically through common ways of seeing the world, or through informal or organizational ties, and that their challenges be contentious in deed as well as in word".

4   Dieter Rucht (1999, 19) argues that modern social movements on the domestic level can no longer be understood as purely anti-systemic, and the reasons offered for this

of production as a basis for economic activity remains normally undisputed by environmentalists. Nevertheless, societal actors demand of states to reform international environmental institutions so that transnational civil society can accept them as forms of transnational political authority that deserve obedience.

Scientific and Technical Service Organizations

The state has traditionally provided a significant part of the infrastructure for scientific research on the domestic level. International cooperation of the sciences or the provision of technical or other forms of assistance for developing countries led to the emergence of various transnational organizations which represent a merger between governmental and non-governmental actors. By mid-2006, the World Conservation Union (IUCN) brought together more than 1000 organizations from 147 countries comprising, inter alia, 84 states, 108 governmental agencies, 749 national NGOs and 82 international NGOs as official members. Service-oriented non-state actors can be recipients of funds from governments or international organizations. They can work as contractors of these institutions, but they can also provide own funds for the services which they provide in regimes. Their services are required during all stages of the political process. The character of their work partially changes during these stages. The level of their involvement normally increases once a regime has become operational.

The services provided by these actors can be classified under three broader categories. First of all, they can contribute to improve the scientific knowledge about the causes and effects of environmental problems or about mid- and long-term effects of different policy options. They actively collaborate in the monitoring of trans-boundary problems, the production of scientific assessments, in the review of the adequacy of commitments, or in reviewing the performance of national policies. The international cooperation among natural or social scientists found expression in the creation of comprehensive non-governmental or semi-governmental science organizations or international research programs. Scientists also contribute to transnational discourse about the long-term effects of environmental pollution. Second, these actors can provide legal expertise for single states, international institutions or other non-state actors. States consult legal experts from non-state actors and include them in delegations to international negotiations or invite them to train members of national administrations. Negotiations in WTO or environmental institutions often raise a broad range of

---

judgment also apply to the transnational level. The emergence of modern social movements no longer signals an existing social order that its challengers would desire its replacement by a more or less fundamentally different type of social order. The decline of political ideologies which aim at change on the level of society as a whole as well as functional differentiation of complex societies transformed domestic social movements to issue-related projects which are interested in achieving improvements in the character of existing institutions.

legal questions. The management of these issues requires specialized legal skills which are often lacking in developing countries. These countries partly rely on the support of environmental service organizations.

Third, service-oriented actors can provide management services for international regimes or support developing countries during the implementation of international policies. IUCN plays a prominent role as that particular non-governmental organization which hosts the independent secretariat of the 1972 Ramsar Convention on Wetlands in its headquarter in Geneva (Bowman 1995, 36–8). Developing countries or countries with economies in transition are often lacking technical, economic, or administrative capacities. Non-state actors can provide various kinds of services during implementation in these countries. These services are often carried out together with programs established in regimes with the aim of supporting implementation in countries which lack capacities.

Multinational Firms and Business Associations

Virginia Haufler (1997, 317) concluded a decade ago, that the "business community plays a role in generating behavioral rules and preferences within regimes, but its role in inter-state negotiations has been somewhat obscured". This has changed during the last decade. Students of global governance recognized the important role of economic actors. Global institutions made efforts to include economic actors in the implementation of their policies. Governments get advice from national or transnational firms or industrial associations and they include representatives from both economic and environmental actors in delegations to international negotiations. Consumers, NGOs, trade unions and policymakers have increasingly demanded from the economy to consider environmental protection, social rights and human rights in their economic activities. Issues of corporate social responsibility have become more relevant in private firms in recent years (Crane, McWilliams, Matten, Moon and Siegel 2008; Rittberger and Nettesheim 2008; Wolf 2008). Economic experts or engineers contribute their knowledge to assessments of regimes about the economic or technical feasibility of policy options. Firms often collaborate with international institutions to support the implementation of international environmental agreements in developing countries.

The divide between economic and environmental interests can lead to severe political conflict in environmental regimes. Such conflict can also determine relations between economic actors. Economic actors consider ecological modernization frequently as a threat for their existence. But many multinational corporations became more open-minded about a contribution by the economy to the achievement of environmental goals. Multinational firms realized that they have to improve their discursive power on the transnational level vis-à-vis noncommercial actors. The establishment of the World Business Council on Sustainable Development (WBCSD) in the mid-1990s remarks one of the most notable efforts in this regard. The World Economic Forum is another actor that influences transnational public discourse. Industrial associations and multinational or national corporations engage in environmental discourse of regimes. Their economic interest can lead them to a behavior which intends to avoid, water down and delay international environmental regulation. But there can also be different economic interests. The expectation that new pro-environmental policies will improve market conditions for technologically advanced products can lead actors from the economy to support the development of international policies. Private firms or industrial associations intensively follow, assess or influence transnational public debate on environmental issues that are relevant for their own economic activities. They make strong efforts to change the framing of environmental issues on the transnational agenda in such a way that economic concerns will be considered adequately. They also initiate discourse within different segments of the global market about the desirability of international environmental policies. This discourse intends to influence policymakers, but it also intends to convince other multinational corporations or industrial associations of the advantages which can arise from a shift to less ecologically harmful ways of production.

International Organizations

States consider international organizations primarily as instruments through which they can achieve common goals. International organizations are "deliberately designed by their founders to 'solve problems' that require collaborative action for a solution" (Ernst B. Haas 1990, 2). International organizations co-ordinate national scientific and technical programs or provide certain functions which could less effectively be managed by independent state action. International environmental organizations contribute to promote learning about cause-effect relationships and to the reformulation of political priorities in international environmental politics. By enhancing the collection, analysis, exchange and dissemination of various kinds of information about environmental problems, international organizations can provide favorable conditions for the formation or further evolution of environmental regimes. International environmental organizations function as arenas which facilitate the articulation of state interests in international politics or the aggregation of various interest groups on the transnational level. They are talk shops within which various actors of world society engage in discourse on problems which require political management or might become relevant for political decision-makers in the future. They are used by states for the establishment of bodies that negotiate international environmental regimes or provide administrative services for international negotiations. During implementation, they can act as effective process managers that provide various services, contribute to capacity building on the domestic level, enable dispute settlement, or support members to achieve effective levels of compliance with regime rules (Breitmeier 1997, Peterson 1997).

Regime analysis has acknowledged that normative institutional factors, such as international organizations, or norms and rules "carry some explanatory weight" for regime formation. It considers international organizations "as supportive rather than as prime causes of regime formation" (List and Rittberger 1992, 102–3). There is no doubt that states have a strong influence on international organizations. States decide about the budget of international organizations. They also influence the selection of staff members or develop guidelines for the policies carried out by IGOs. But international organizations can also take a more autonomous role in the political process. International expert groups can influence political agendas and develop programs whose contents elude the constraints established by national policy-makers. States often respond to ideas, to the rise of new issues on the international political agenda, or to policy solutions which emerge from the work of international organizations. Partly independent action of international organizations may be desired by states when divergent interests avoid that consensus can be achieved about the character of policies.[5] The debate about a reform of the global trade regime reveals that there can be strong differences among international organizations concerning the relevance which they ascribe to environmental concerns in implementing their functional goals. The functional character and corporate identity of various international organizations partly determine how strongly they will refuse or support the development of international environmental policies.

Individuals

The role of individual leadership for the establishment or further evolution of international institutions has long been neglected by students of world politics. The emergence of institutional order on the global level is closely linked with the role of single individuals who raise public attention for single issues, try to convince decision-makers about the salience of trans-boundary problems and the need for institution-building, come up with compromise solutions which overcome gridlock in negotiation processes, or take the initiative for transnational self-regulation.[6]

In many cases it is insufficient to perceive of an individual's contribution to global policymaking exclusively as an expression of the will of a state, an

---

5 Many examples could be offered for such a more autonomous role of international organizations or programs in global environmental governance. For the case of international environmental protection of the Mediterranean, Jorgen Wettestad (1999, 27) argues that "it seems as if the active role of the United Nations Environment Programme (UNEP) was more or less a necessary condition for making the cooperation work in the initial phases" and that the organization "was also an active player in the political process, sponsoring workshops, among other things".

6 Martha Finnemore (1999, 163) illustrates the significance of "a few morally committed private individuals – individuals without government positions or political power – and the elite networks they were able to use" for the foundation of the ICRC and the Geneva Convention. Further on, the author points out that "neither of these has received much attention from international organizations scholars" and that the "state-centric analysis dominating the field has made it difficult to recognize that private individuals with no formal political standing might have significant influence".

international organization, or a non-state actor. Of course, individuals are mostly representing the interests of those collective actors to which they owe their leadership roles. Individual leaders in global policymaking are also facing the challenge to convince their role-givers of the salience of a trans-boundary problem, of the need for the introduction or readjustment of international programs, or of the feasibility of policies discussed in the context of international negotiation systems. It would be one-sided to consider individual leadership in world politics only as a product of the preferences of states, international organizations or non-state actors as role-givers. Individuals themselves can influence the interests of collective actors to which they are responsible. Individuals can be policy entrepreneurs on the transnational level and make efforts to set an issue on the transnational agenda, formulate policy solutions, or use windows of opportunity to promote their political projects in global governance systems. The role of individuals in world politics also poses critical questions: Does the charisma or the expertise possessed by individuals already suffice to consider them as forces which can contribute to improve the legitimacy of global governance? The expansion of policymaking from the domestic to the global level runs the risk of empowering transnational elites composed of individuals who share common understandings about the kind of policies which they wish to implement on the global level. Collective associations like states, international institutions or non-state actors will have to establish procedures for the selection of these experts or find ways through which they can guarantee that individual leaders will be accountable to the role-givers which they represent. Whether the goals which these actors pursue conform to those that exist in the broader transnational public can be reviewed if regimes will offer procedures that allow exploring the activities of these actors.

Transnational Networks

Networking can reduce transaction costs and increase the flow of information between like-minded transnational actors.[7] It intends to increase the influence of these actors on the transnational agenda and to produce strategic advantages vis-a-vis other actors in world politics. It can be used to influence the implementation of policies or programs. New communication technologies have improved the conditions for non-state actors for the transnational exchange of information and for discussing and coordinating their political initiatives or the scientific and technical work which they carry out in international regimes. The existence of new channels of communication alone can not sufficiently explain the rising influence of transnational networks in world politics. Robert O. Keohane and Joseph S.

---

7   For the level of domestic politics, such policy networks were conceived of as "new forms of political government which reflect changed relationships between state and society" and which can be used as "mechanisms of political resource mobilization in situations where the capacity for decision-making, program formulation and implementation is widely distributed or dispersed among private and public actors" (Kenis and Schneider 1991, 41).

Nye, Jr. (1998, 90) argued that in times of massive expansion of communication "political struggles focus less on control over the ability to transmit information than over the creation and destruction of credibility". Thus, advocacy networks or transnational networks of technical and scientific experts will only be able to influence international policy-making if they disseminate credible information.

Various types of networks emerged in global environmental politics. These networks differ with respect to the density of interactions among members, the range of actors which they include, the durability of their existence, their formal and informal character, or the degree to which they collaborate with states. From the perspective of nonprofit activist environmental groups, Ingo Take (2002, 142–5) illustrates the emergence of transnational alliances within the world of nonprofit activist groups or between these environmental activists and like-minded commercial actors or states. The world of nonprofit-oriented non-state actors uses transnational networking for enhancing the exchange of information, for coordinating political action, or for the development of joint political positions. Networks among environmental groups can be charged with tension. These actors can compete with one another, disagree about political strategies, or have different identities as advocacy or service organizations. These conflicts can occur within networks as well as between them. Nevertheless, these actors often complement one another when the pressure of advocacy organizations gives service-oriented non-state actors political backing for their contribution to the management of international regimes. Transnational networking occurs also outside the world of nonprofit activist environmental groups. Networks among transnational economic actors frequently represent the interests of the international business community. Finally, policy networks between state actors and non-state actors became increasingly relevant for designing or implementing policies in international environmental regimes.

Institutional Design, Non-State Actors and the Legitimacy of Regimes

Political analysts frequently argued that the management of complex problems requires a transition from a hierarchical to a horizontal mode of governance. This change in the mode of governance has first occurred in industrialized countries. Denationalization and the broad range of actors affected by policymaking made it necessary to include societal groups in policymaking in order to guarantee effective implementation of international policies (Scharpf 2000, 338–51). Our prime hypothesis suggests that a similar transition will be required also on the level beyond the nation-state. So far, we are lacking sufficient knowledge whether the empirical reality in global governance systems is determined by such a development. Approaches that pin their hopes on transnational civil society assume that political power or capabilities of non-state actors will lead to the achievement of more effective or democratic international institutions. Studies which deal with advocacy-networks or social movements imply that participation, contestation, or civil disobedience will cause improvements in the outcomes of international institutions, because governments are forced to respond to the political demands of protesters.[8] Movement-oriented approaches conceptualize regimes and transnational actors as two separate spheres. If hierarchical forms of governance change towards a more horizontal mode, such an originally strict division between the spheres of regimes and transnational actors dissolves. The notion of a more horizontal mode of governance implies that social groups influence institutions not only through external pressure and contestation but also through their contributions to the development and implementation of policies.

Our prime hypothesis will be explored in two subsequent steps. First of all, we need empirical proof of the assumption that institutional mechanisms can lead to changes in the preferences and behavior of states. We must then examine the roles played by non-state actors in these mechanisms. Consequently, two explanatory approaches will be used for the empirical analysis of the database. The institution-based explanation refers to the broad strand of regime literature which explains regime effectiveness as a result of the institutional design. The approach which focuses on non-state actors takes into account that these actors can impact on regimes in various ways. It will consider activities of these actors during agenda-setting, negotiations, or implementation.

Institutional Mechanisms and Legitimacy

Whether international regimes contribute to changing the preferences and behavior of states is still a question that is highly disputed between theorists. Some energy has been spent by institutionalists on identifying institutional design principles which can be considered as primary causes for the production of effective outcomes on various levels of policymaking (March and Olsen 1989; Goodin 1996; Ostrom 1990).[9] A first effort to explore the impact of institutional mechanisms of regimes on outcomes and impacts on the basis of a broader set of cases was made by a collaborative project between Norwegian and North-American scholars. This group considered regime effectiveness as a function of the character of the political

---

8   For example, it has been argued that "it is increasingly clear that transnational social movements influence the outcomes of international relations at least by interacting in and shaping the *political processes* that generate global policy" [*emphasis by author*] (Smith, Pagnucco and Chatfield 1997, 74).

9   For example, Elinor Ostrom (1990) explored the impact of design principles on the effective management of small-scale common-pool resources (CPRs). A design principle is conceived of as "an essential element or condition that helps to account for the success of these institutions in sustaining the CPRs and gaining the compliance of generation after generation of appropriators to the rules in use" (Ostrom 1990, 90). Ostrom's work suggests that the existence or non-existence of various institutional design principles had an impact on the evolution of robust, fragile, and failed institutions. Only those of the explored CPRs which possessed the whole range of design principles considered relevant by the author have been governed by robust institutions (Ostrom 1990, 180).

problem and the problem-solving capacity (Miles, Underdal, Andresen, Wettestad, Skjaerseth and Carlin 2002). Following a game-theoretic logic, the project distinguished "*malignancy*" and "*benignity*" as two characteristics of a political problem. Malign problems were perceived of "primarily as a function of the configuration of actor interests and preferences that it generates", whereas benign problems were characterized by identical preferences (Underdal 2002a, 15). Such types of social problems have been distinguished by various situation- or problem-structural approaches. Each of these approaches correlates the characteristics of a problematic social situation or problem structure with the evolution of cooperative outcomes or the effectiveness of institutions.[10] The types of predominantly *benign* and predominantly strongly *malign* problems are two extremes on a range of four types of problems. Malign problems are perceived of as "problems of incongruity". Their essence consists in the distortion of incentives, because the costs or benefits that arise in connection with a regime broadly vary among actors. When actors are located in a context which is mainly characterized by competition, they are particularly susceptible to assessing their own costs and benefits in relative terms. Additional factors like the degree of asymmetry that exists between states (e.g., the upstream-downstream relationship or the incompatibility of values) are considered to further determine the malignancy of a problem.

The Norwegian-American project explored a set of fourteen environmental regimes. In a summary of the project's empirical findings, Aril Underdal (2002b, 435) concluded that regimes "can have a fair amount of success in dealing also with malign problems". Further on, the author points out that in those cases where "the problem that motivated the establishment of the regime has become less severe, we find that the regime has been as important or more important than other factors in more than half of the cases". The project revealed that the interplay between politically malignant problems and strong uncertainty in the knowledge base finally involved that problems remained intractable. The empirical findings of the project confirmed that effective management of politically malign problems requires the provision of a significant amount of problem-solving capacities, of selective incentives, or the creation of linkages with less complex issues. In addition, the findings suggest that the provision of a sufficient level of skills and energy needed for problem-solving improve the conditions for political learning and also improve the conditions for the effective management of politically malign problems.

The exploration of the contribution of institutional mechanisms involves illustrating those causal mechanisms through which various grounds for legitimacy can be achieved. This approach is based on the assumption that various uncertainties in the cognitive setting can be reduced by the provision of international programs. These programs alleviate the collection and exchange of information about the causes of a problem, contribute to the development of policy options or improve

---

10 On these approaches consult Efinger, Rittberger and Zürn (1988); Hisschemöller and Gupta (1999); Wettestad (1999); Young (1999a, 50–78); Zürn (1992).

capacities to participate in social practices in regimes. The demand for these design principles has been emphasized by students of regime effectiveness. They explored the functionalist role of regimes or their contribution to improving the knowledge base (Haas, Keohane and Levy 1993). In addition, institutional mechanisms are assumed to improve compliance by states by reducing the incentives for cheating, by reducing uncertainties on the part of a state about compliance by others, or by improving capacity-building for domestic implementation. The study of compliance is determined by ongoing debate regarding whether this ground for legitimacy can be achieved by the use of a management or enforcement approach (Chayes and Mitchell 1998; Downs, Rocke and Barsoom 1996).

Fulfillment of various other grounds for legitimacy can be a precondition for goal-attainment or problem-solving. Thus, an institution-based explanation for goal-attainment or problem-solving must explore how far impacts of institutional mechanisms on the knowledge-base or on compliance are reflected in observed levels of goal-attainment or problem-solving. This approach also suggests that institutional mechanisms can be used for achieving a more equitable distribution of costs and benefits among regime members. In this regard, Oran R. Young's (1994) approach on institutional bargaining has taken into account that self-interest alone will not suffice for the evolution of international regimes. It emphasizes equity of the arrangements that result from negotiations or procedural fairness as two among a number of necessary conditions, because "institutional bargaining in international society can succeed only when all the major parties and interest groups come away with a sense that their primary concerns have been treated fairly" (Young 1994, 109). The provision of programs which reduce asymmetries of distributional outcomes between regime members can be a precondition that must be fulfilled in specific contexts of world politics before single countries are willing to accede. The provision of institutional mechanisms for financial or technological transfer or other forms of assistance can be assumed to contribute to changing governmental and societal preferences in developing countries or in countries with economies in transition towards more regime-conducive attitudes.

Non-State Actors and Legitimacy

The functioning of institutional mechanisms relies on social actors. While international regimes have traditionally been understood as state-dominated forms of governance, our prime hypothesis leads us to explore whether the relevance of non-state actors in regimes increased over time. A focus of analysis on the role of non-state actors in the management of regime functions suggests that this approach can complement the institution-based approach. This approach considers that non-state actors can have diverging preferences regarding the evolution of regime policies. The approach starts from the assumption that non-state actors can have a positive impact on the fulfillment of various grounds for legitimacy in world politics. It assumes that conflicting interests on the transnational level will not inevitably lead to policy gridlock in the long-term. This approach assumes that non-state actors contribute to exploring the causes and effects of environmental problems, to monitoring the evolution of environmental pollutants, to compliance monitoring or to reviewing the adequacy of commitments, or to the implementation of international policies. While participation in the management of programmatic activities is considered as an important contribution of non-state actors, their political activities in the transnational public sphere are of equal importance for this approach. This approach considers the activities of advocacy or movement-oriented actors as well as of those non-state actors which are participating in global policy networks.[11] The approach implies the assumption that emerging global civil society practices a division of labor. Activist environmental groups focus on the politicization of environmental issues in the transnational public sphere. Other actors collaborate with states in the management of regime policies. Those groups which correspond to activist environmental non-state actors primarily focus on pressurizing states to speed up the creation or further evolution of global governance systems. Non-state actors which are embedded in networks benefit from the improvement of participatory procedures or from the expansion of international governance. It could be argued against this assumption that the conflict between pro-environmental and economic interest groups occasionally blocks the achievement of effective solutions in world politics. It can not be excluded that an insurmountable divide occasionally characterizes relations between these interest groups. However, the behavior and policies of private actors in world politics must consider the social reason that determines transnational public debate. These actors have to justify their policies and behavior vis-à-vis the transnational public sphere. Since they are involved in rational discourse about the appropriateness and about the possible consequences of their policies, they are normally faced with the challenge of developing positions which converge to social reason.

The role which non-state actors play for the fulfillment of grounds for legitimacy is dependent on the function possessed by these actors during the different stages of the policy process. In exploring the role of commercial and noncommercial non-state actors, three stages of the political process will be distinguished: agenda-setting, negotiations, and implementation. These stages often overlap or run parallel. For example, agenda-setting not fully ends with the beginning of formal negotiations. The political pressure occurring outside negotiations is just as relevant as the politicization or framing of issues during agenda-setting. Environmental actors must enforce their discursive power vis-à-

---

11   Much of the scientific literature dealing with the working of international institutions focused on the contributions of various non-state actors to the effective management of regime functions. Wolfgang H. Reinicke (1998, 90) proposes the use of "public-private partnerships" in global governance systems which will "generate greater acceptability and legitimacy for global public policy". This argument is derived from the insight that non-state actors have a direct stake in the output, outcomes and impacts of public policy and that they can improve "information, knowledge and understanding of increasingly complex, technology-driven, and fast-changing public policy-issues".

vis opponents during the stage of agenda-setting. They intend to achieve that the environmental issues for which they lobby "will be considered legitimate items of public controversy" (Cobb and Elder 1972, 163). Agenda-setting is determined by competition between various types of non-state actors, but also between transnational societal actors and states. Environmental non-state actors must achieve that an environmental issue will triumph over competing issues in this process of issue selection. Pro-environmental non-state actors make efforts to gain supremacy in the fight for discursive hegemony vis-à-vis those interest groups or states which have opposed interests (Kriesberg 1997, 17–18). The activities of noncommercial and commercial non-state actors during agenda-setting can be directed on a broad variety of topics. For example, discursive power can be used to raise public awareness for a problem, to direct public attention to new consensual knowledge, to support consideration of policy options, or to politicize the need for a more equitable distribution for costs and benefits.

During negotiations, states can be confronted with a number of serious difficulties which constrain the achievement of solutions for problem-solving. Negotiations can be determined by various kinds of uncertainties: knowledge about cause-effect relationships can be so insufficient that states fail to agree on joint international policies; the lack of suitable policy options can delay negotiations; uncertainty on whether other states will comply or implement international policies may involve that single states will refuse to accede to an agreement; asymmetries between negotiating states regarding the availability of capacities for problem-solving or which affect the distribution of possible outcomes can be so severe that negotiations fail to reach agreement (Scharpf 2000, 199f). This approach takes into account that non-state actors can become members of national delegations or advisors to negotiating bodies where they contribute to reducing uncertainties. But one has to be aware that states are dominating negotiations and that non-state actors which participate in national delegations to negotiations have to fall into line with the interest of their national governments. The contributions which non-state actors can make during this stage partly depend on the open-mindedness of states about such expertise or on the rules of procedure which govern participation by non-state actors in international negotiation systems. The phase which covers the operationalization and implementation of regime rules begins with steps like ratification or the formation of regime bodies which are required by states to make international agreements operational. Processes of domestic ratification can be determined by more or less strong conflict over the terms of an international legal agreement. Environmental non-state actors can pressurize states to ratify. In doing so, they must counterbalance interest groups which intend to delay or prevent ratification or implementation.

Chapter 5
Regimes, Case-Design, and Coding Procedure

International regimes are social institutions in issue-areas which consist of agreed upon principles, norms, rules, decision-making procedures and programmatic activities. They are based on formal international legal agreements, but they can also consist of informal agreements or social conventions. The IRD-project selected a set of "hard law"-regimes for coding. Each of them has been based on one or various legally binding international agreements. In a number of cases, "soft-law"-regulations complemented "hard-law"-provisions of regimes. Before the coding of a regime could begin, a number of conceptual issues had to be clarified. These issues center around one basic question: What is a case? The signing of an international legal document (whether it has been legally binding or not) can be considered as the date that marks the beginning of a regime's existence. But regimes can also expire. The topic of regime demise has been ignored by regime analysts.[1] The research community paid far more attention to exploring the creation or further evolution of international regimes. Only such regimes were included in the database which were in operation until the end of the twentieth century. The year 1998 represents the end point of the life-cycle explored for each regime. The coding has been carried out from 1999 onward.

By focusing analysis on cases which share identical temporal end points, it will be possible to assess how far grounds for legitimacy have been fulfilled by global environmental governance systems at the end of the twentieth century. Some environmental regimes had historic forerunners which were later followed by new institutions. The identification of the starting point of a regime has been guided by the conceptual decision that an institutional predecessor will be disregarded, if the operation of this earlier regime has been discontinued. A case in point is the regime for the international regulation of whaling where a first convention had been signed in 1931. This convention was supplemented in 1937. But important whaling states remained non-signatories for some time and no concrete measures were established in that regime. During World War II, the regime was *de-facto* not in operation. Thus, the creation of the International Convention for the Regulation of Whaling of 1946 is considered as the starting point for the current whaling regime.

A similar problem concerning the delimitation of a temporal starting point for a regime arises for those cases where regimes are nested in broader institutional

---

1   Surprisingly, this topic has also been neglected by those analysts who, like neo-realists, raise serious doubts as to whether regimes could survive major challenges arising from developments within or outside an issue area.

settings that were established before the actual environmental regime emerged. The regime for environmental management of the Great Lakes provides a good example for regime nesting. The Boundary Waters Treaty concluded between the United States and Canada in 1909 primarily focused on regulating conflicts over the use of water between both countries. Thus, it also applied to the area of the Great Lakes. The International Joint Commission (IJC) established by this treaty has been given various functions to fulfill the goals of the Boundary Waters Treaty. Both states agreed in Article IV of the treaty to prevent pollution of boundary waters. But only the Great Lakes Water Quality Agreements of 1972 and 1978 established a regime for the environmental management of the Great Lakes. This regime partly emerged as a result of IJC's work in the 1960s and early 1970s. The work of the IJC revealed that political management should pay more attention to the issue of water pollution in the region of the Great Lakes (Valiante, Muldoon and Botts 1997, 199–203).

A number of other questions had to be resolved during the development of case structures: How can we delimit a regime from other institutions? How can we cope with the fact that more than one legal agreement normally constitutes a regime? Which criteria can be used for distinguishing different time periods in a regime? These questions have been resolved with coders during so-called "precoding negotiations". These negotiations were carried out before the actual coding of regimes began. The development of case structures emerged from intensive discussion with coding experts. These discussions were based on criteria for the development of a case design that were specified by the project team (Breitmeier, Levy, Young and Zürn 1996a and 1996b). In the following, the conceptual issues which were discussed with case study experts during pre-coding negotiations will be illustrated in detail.

International regimes are problem-driven arrangements which are established for the management of trans-boundary problems in an issue area. Thus, a first step will describe various characteristics which determine the framing of transnational environmental problems. In a second step, international regimes will be considered as institutions which are based on legal agreements.[2] These legal arrangements can address a variety of sources of environmental damage, aim at different regulatory targets, or establish different mechanisms necessary for effective problem-solving. A regime will be reduced to its legal component parts or be subdivided in different time periods. The smallest unit of analysis will be the so-called "regime element". It consists of a legal agreement or a number of legal arrangements that exist during a distinct time period. Finally, the cases that were chosen for coding will be described.

---

2 The texts of international legal agreements that will be referred to in the empirical chapters of the book are available in various volumes (Weiss, Magraw and Szasz 1999, UNEP 1996) so that quotations for individual agreements refer to these volumes unless otherwise noted.

International Regimes as Problem-Driven Institutions

One of a number of grounds for the legitimacy of international regimes refers to the contribution made by the institution to problem-solving. How are "problems" defined in global environmental politics? The framing of a problem can involve various aspects: the activities that exist in the social world causing the problem; the effects caused by these activities on the natural environment; measures that can be chosen for improving the state of the environment; the costs of problem solving; or different preferences of states regarding solutions or policy options. A problem is perceived as a collection of various phenomena that can be observed in the natural or social world. Frequently, it is impossible to identify a problem unambiguously because a regime addresses more than one problem. Problems are normally more clearly defined than the broader issue-area. The borders of issue-areas are often vague and blurred. But the problems managed by an institution normally involve the basic objects of contention (e.g., the environmental good or natural resource managed by an institution) which are considered for political management. They can also involve constellations of diverging interests existing among actors.

States, international organizations, or domestic and transnational actors participate in the framing of problems. They develop publicly shared understandings on basic issues like the causes and effects of a problem, about what should be maximized in the issue-area, or about the political options that exist for problem-solving.[3] The framing of problems is a constant process. This process can not be limited to the stages of agenda-setting or to the negotiation stage alone. International institutions can experience a re-framing of certain problems, or can be confronted with the emergence of new problems that call for political management. They can also experience a fundamental change that leads to a re-definition of the problem as a whole. A cross-issue comparison made by The Social Learning Group (2001a and 2001b) between the issues of climate change, ozone depletion, and acid rain showed that the framing of these problems has been subject to severe changes over time. Some problems were re-framed or were determined by a partial shift in emphasis concerning the relevance ascribed to single sub-problems or to approaches for political management (van Eijndhoven, Clark and Jäger 2001).

What is the Problem?

The framing of a problem can be characterized by more or less deep conflict between social actors: Is there a problem at all? What are the causes and effects of the problem? How can the problem be remedied? Framing consists of a process where parties that

---

3   In discussing the influence of social actors on framing processes, Doug McAdam, John D. McCarthy and Mayer N. Zald (1996, 6) define framing on the basis of David Snow's original conception as "conscious and strategic efforts by groups to fashion shared understandings of the world and of themselves that legitimate and motivate collective action".

are causing or that are affected by a problem dispute on whether distinct cause-effect relationships can be identified that justify the consideration of a problem as existent, or whether the problem deserves special attention by policymakers.

Even if participants agree that there is a problem in an issue area, they can be divided over its real character. For example, states and other actors can agree that it will be necessary to develop international norms that govern a particular resource; there can be disagreements between them on whether the conservation or the utilization of a resource should be considered as a starting point for the framing of this problem. If the focus is directed on the management of resource depletion, the framing stresses the environmental character of a problem. However, a focus on the utilization of a resource directs political management on economic aspects. The framing that refers to a problem's environmental character involves the development of conservation goals, whereas a predominantly economic view of the problem entails the design of goals that primarily refer to the economic use, trade, or distributional aspects that arise in connection with the exploitation of the resource. This antagonism can often not fully be resolved in regimes. Negotiations are often determined by efforts to balance economic and conservationist interests. The international management of issues like whaling or trade of endangered species has been determined by continuous conflict between regime members on how they should define the problems that are managed by these regimes (Andresen 2002, 383ff; Sand 1997, 181; Zangl 1999), since economic interests (e.g., of whaling companies) compete with conservationist interests.

Geographical Scope and Transnational Identities

The broad majority of environmental regimes has not a global scope. Many regimes have been established for the management of bilateral, multilateral or regional problems. The geographical scope of a problem is often self-defining. This applies to regional seas, inland seas, or trans-boundary rivers. But in some of these cases it can be necessary to delimit the issue-area geographically from neighboring ecosystems. Single problem types like the management and conservation of common pool resources that can heavily be fenced (e.g., fish stocks) in some way elude geographical delimitation. The framing of a problem's spatial domain can also expand in the course of time as a result of a geographical expansion of the causes and impacts of a problem. The management of a problem can be delegated to different regional contexts of policymaking. For example, the regional annexes established in the desertification convention illustrate that a focus of political management on the global level alone will not suffice. A global framing is often complemented by additional regional framings of sub-problems that exist below the global level.

The Interplay of Problems and Between Institutional Contexts

Since ecosystems overlap, it is normally difficult to exactly delineate the boundaries of environmental issue-areas. Similar overlaps exist between different environmental problems. Ozone-depleting substances contribute to the warming of the climate because of their greenhouse potential. Significant reductions in emissions of these substances cause positive impacts on the climate system.

Meanwhile, regime analysts also pay attention to the interplay of problems and to the institutional linkages which have been established between regimes, international organizations, or aid programs for improving the quality of problem-solving (Oberthür 2001; Oberthür and Gehring 2006; Young 2002b). The member countries of the Black Sea regime can not prevent the discharge of pollution resulting from upstream countries of the Danube which flows into the Black Sea (Bonacci 2000; Kaspar 1999; Linnerooth-Bayer and Murcott 1996). While the so-called Bucharest Convention on the Protection of the Black Sea against Pollution of 1992 notes "that pollution of the marine environment of Black Sea also emanates from land-based sources in other countries of Europe, mainly through rivers", only the six countries that are bordering the Black Sea have signed and ratified this document by mid 2002. But important upstream countries of the Danube which discharge harmful substances into the river have so far not taken steps to become members of this regime. International aid agencies launched a multiyear program to reverse nutrient over-enrichment and toxic contamination of the Danube/Black Sea basin. These measures also intend to improve the ecological conditions in the Black Sea.[4]

The framing of environmental problems often considers that some of the causes can be produced by economic institutions. Environmentalists strengthened complaints about the implementation of programs which were designed by the World Bank and by other international organizations to promote the setting-up of industry and the industrial use of agriculture or which were aimed at improving the infrastructure in developing countries. Because these policies frequently failed to consider and avoid negative effects of these international policies on ecosystems, they provoked strong criticisms of environmental NGOs and indigenous people which were often negatively affected by projects financed by international programs.

Regime Dynamics, Institutional Setting, Important Actors

So far, our discussion led to a number of criteria which govern the development of a case structure: a regime rests on one or more problems; it has a clearly-defined temporal starting point beginning with the signing of an international legal agreement; it can be delimited from other institutions or be separated

---

4 Another example illustrates, that possible contributors to environmental problems that are located outside the issue-area can become official members. Switzerland and Luxembourg as land locked countries that are some hundred kilometers away from the North Sea coast are members of the OSPAR-Convention. These countries are bordering on the Rhine which discharges harmful effluents into the North Sea (Skjaerseth 2000).

from legal settings in which it is nested; regimes refer to a more or less clearly delimited geographical scope. Two further questions arise in connection with the development of a case structure: How can the variety of legal agreements that often constitute a regime be reflected in a case structure? How can a case structure consider the political or institutional dynamics which determine the further evolution of international regimes? In some respects, both questions are interrelated. The dynamics that determine the evolution of international regimes often finds expression in the expansion or complete revision of existing international legal agreements. The moment of a watershed when a regime experiences a significant change in the composition or character of legal agreements can be used for distinguishing different time periods during the evolution of a regime. In the following, it will first be described how a case structure emerges from various legal agreements that constitute a regime. In a second step, attention will be given to those factors that can be used to identify watersheds which separate different time periods in regimes. These steps illustrate that regimes can be subdivided into various regime components and that different time periods can be distinguished for these components. Finally, it will be illustrated which important actors have been selected in pre-coding negotiations. Regimes are not only understood as problem-driven arrangements that are based on various types of legally binding or non-binding agreements. But they are also perceived of as institutions that emerge from the activities of various social actors.

Legal Agreements and the Case Structure of a Regime

The number of legal agreements or their function for the management of single problems determine whether a case will consist of separate regime components. A few regimes consist of only one component. A case in point is the tropical timber trade regime that has been established with the International Tropical Timber Agreement in 1983. This agreement has been replaced by the 1994 International Tropical Timber Agreement which was similar in most respects to the former agreement and which included no substantive changes in the normative or procedural arrangements of the regime (Gale 1998). Apart from the existing legal agreement that constitutes a regime, separate regime components have been identified in the process of developing a case structure if distinct institutional forms like separate treaties, protocols, annexes, amendments or non-binding programs emerged in a regime. In many cases, the legal development of environmental regimes has been characterized by a "framework convention/protocol-approach". This approach implies that institutionalization begins with the negotiation of a general legal framework that provides norms about the institutional set-up or the procedures used for decision-making before it will be followed by other legal documents that provide for concrete environmental policies. Protocols, annexes, or amendments are added to framework conventions to specify the reduction of various forms of pollutants or the prohibition of other environmentally damaging practices. Many regimes included in the database can be referred to as examples which illustrate that states relied on this approach. The regimes addressing global issues like the protection of the ozone-layer, climate change, the conservation of biodiversity, or desertification are based on framework-conventions.

The creation of a new legal arrangement does not inevitably result in the development of a separate regime component. Such automatism would lead to the overexpansion of case structures for single regimes and produce unreasonable workload for coding case study experts. It would also imply that each legal document that has followed a framework convention or other type of legal document constituting a regime must be given equal relevance. The analysis of the legal frameworks that constitute regimes often reveals that not all legal arrangements belonging to a regime lead to the creation of new regime components. At least one of the following conditions had to be fulfilled before a legal arrangement was given the character of a regime component. First, these distinct institutional forms had to address a sub-problem that can be separated from the problem that has already been addressed by the regime. No legal agreement met this criterion due to the fact that the problems addressed by single regimes were rather broadly defined so that subsequent legal agreements could always be understood as responses to the initial problem. Second, if these additional institutional arrangements covered differentiable sources of the problem(s), they were treated as a basis for separate regime components. Many of the additional components of regimes emerged because separate legal arrangements were designed for the management of various sources of a problem. For example, the ozone-regime established with the Vienna Convention in 1985 includes a number of institutional arrangements. The Montreal Protocol of 1987 or its adjustments and amendments agreed upon in London (1990) or Copenhagen (1992) have been treated as different components of this regime because they manage the reduction or phase-out of various ozone-depleting substances. Third, separate components have been identified if legal agreements were aimed at different regulatory targets. The Antarctic regime has been perceived of as a collection of different legal arrangements which, apart from the 1959 Antarctic Treaty, address different regulatory targets. Each of these agreements can be considered as a regime component. The Antarctic Treaty represents the regime component that emerged at the outset. Other important legal agreements such as the 1964 Convention for the Conservation of Flora and Fauna, the 1972 Convention for the Conservation of Seals, the 1980 Convention on the Conservation of Antarctic Marine Living Resources (CCAMLR) complement the Antarctic regime together with the 1991 Protocol on Environmental Protection. The fourth condition specifies that a separate component will be identified if legal agreements aim at different clusters of actors. It takes into account that regimes can treat various groups of actors differently. This condition did not have an effect for the identification of regime components. Though the Antarctic Treaty distinguishes between Consultative Parties and non-Consultative Parties, the condition did not apply in this case. Because the treaty itself represents the original founding document of the Antarctic regime, it has been considered as a regime component that constitutes the regime. Even though some institutions like the ozone regime include provisions which distinguish between developed and developing countries, the legal agreements established in these regimes have been perceived of in terms of their contribution to the reduction of single sources of the environmental problem rather than with respect to the distinction between different clusters of actors. Finally, in some regimes components have been distinguished with reference to the fifth condition after which institutional forms can deal with major regime functions. The climate regime's Financial Mechanism agreed upon in UNFCCC, the ozone regime's Multilateral Fund, or the compliance mechanism established in the regime on fisheries management for the South Pacific region represent such regime functions.

The decision that the coding will focus on regimes that rest on hard law rather than on soft law provisions has been made for a number of reasons. The debate about the contribution which institutions can make for the evolution of social order in world politics has primarily focused on hard law regimes. The more recent attention given by students of international governance to various types of soft law regimes will improve our understanding on whether less institutionalized forms of governance can contribute to the evolution of social order. In some respects, the distinction between hard law and soft law regimes is artificial. Many international regimes based on hard law emerged from non-binding legal arrangements which were further developed and transformed by states into institutions with binding obligations. Regimes can often involve a mix between hard law agreements and soft law provisions. Components of some regimes included in the IRD are based on soft law regulations which complement hard law regulations. The regime that emerged with the 1949 Convention for the Establishment of an Inter-American Tropical Tuna Commission has been completed by a component for the conservation and management of dolphins in that region in 1976 which is based on soft-law provisions (Joseph 1994, 4–5). The "Rhine Action Programme" of 1987 as another soft law document set the course for intensifying inter-state collaboration for the protection of the Rhine. The introduction of soft-law-components into existing hard law regimes has frequently been practiced and both types of regulation must not inevitably contradict with one another.

Components and Watersheds: Regime Elements as Smallest Units of Analysis

Some regimes came into existence several decades ago. The empirical study of regimes raises the question on whether changes in the institutional framework had an impact on the outcomes of these regimes. Identifying the impact of such changes on the fulfillment of grounds for legitimacy will require subdividing regime components into different time periods. The smallest unit of analysis – the so-called "regime element" – consists of a regime component that is explored for a distinct time period. From an analytic perspective such a distinction between time periods is helpful. But such distinction between separate time periods can not be made for each regime. It depends on whether single conditions are fulfilled that allow to subdivide the process of regime evolution in more than one time period.

Such distinction is based on the concept of watersheds which involves a number of conditions of which at least one must prevail so that regime components can be divided into separate time periods. One watershed has been determined for 13 regimes included in the IRD. Two watersheds have been identified for three regimes. Only seven regimes have not experienced a watershed. Of the 19 watersheds identified, only three emerged in the 1970s. Seven watersheds occurred in the 1980s, eight in the 1990s and one of them between 1989 and 1991. The frequency of watersheds identified for the past two decades indicates that strong dynamics determined international environmental governance during this period. It found expression in the re-framing or further evolution of existing institutional frameworks dealing with environmental protection. The separation of institutions into various regime components and distinction of watersheds had the effect that the 23 regimes included in the database have been separated in ninety-two regime elements (see Table 5.1).

Which criteria governed the identification of watersheds? Which impacts emerged from these criteria for the development of case structures? First of all, a watershed occurred if a significant restructuring of principles or key norms could be observed in the issue area. For example, the transition from resource management to environmental protection as a central concern of the Antarctic regime made between 1989 and 1991 led to the identification of a watershed for this regime. It finally found expression in the adoption of the 1991 Protocol on Environmental Protection to the Antarctic Treaty. Second, watersheds have been identified if a significant change in the group of leading actors which are members of a regime could be observed. Some watersheds have been determined by such change in the composition of regime membership. For example, only the agreement that could be reached in 1990 from industrialized countries to the establishment of a Multilateral Fund paved the way for accession of India and China to the ozone regime. Further on, the admission of 14 new Consultative Parties which began in the late 1970s and which was continued throughout the 1980s required to distinguish the earlier period of the Antarctic treaty system from a new period that emerged in the 1980s. Third, a watershed also emerged if the functional scope of a regime expanded as a result of new legal agreements that deepened regime rules significantly. There are many examples which illustrate that watersheds have been identified with reference to this condition. The London Amendment to the 1987 Montreal Protocol led coding case study experts to determine the year 1990 as a watershed of the ozone layer regime since only the regulations agreed upon in London involved a significant broadening in the scope of the regime rules.

The distinction between regime components or the identification of watersheds made during the pre-coding negotiations involved that case structures broadly vary concerning the number of so-called regime elements that constitute a case. Five regimes neither consisted of more than one regime component nor did they experience a watershed. The regimes on biodiversity or desertification have emerged only during the first half of the 1990s and the Cartagena Protocol on Biosafety which was added to the biodiversity convention in 2000 lies beyond

Table 5.1   Regimes and Regime Elements

Regime	Watershed(s)	Regime Elements
Antarctic Regime 1959–98	1980 and 1989/91	Antarctic Treaty (1959–80) (1980s) (1989/91–98)
		Conservation of Flora and Fauna (1964–80) (1980s) (1989/91–98)
		Conservation of Seals (1972–80) (1980s) (1989/91–98)
		CCAMLR (1980s) (1989/91–98)
		Protocol on Environmental Protection (1989/91–98)
Baltic Sea Regime 1974–98	1992	Principles of Co-operation (1974–92) (1992–98)
		Environment Protection Principles (1974–92) (1992–98)
		Regulations for all Sources of Marine Pollution (1974–92) (1992–98)
		Nature Conservation (1992–98)
Barents Sea Fisheries Regime 1975–98		Norwegian-Russian Cooperation on Fisheries in the Barents Sea Region (1975–98)
Biodiversity Regime 1992–98		Convention on Biological Diversity (1992–98)
CITES-Regime (Trade in Endangered Species) 1973–98	1989	CITES-Convention (1973–89) (1989–98)
		TRAFFIC-Network on Monitoring and Compliance (1978–89) (1989–98)
Climate Change Regime 1992–98	1997	United Nations Framework Convention on Climate Change (1992–97) (1997–98)
		UNFCCC Financial Mechanism (1992–97) (1997–98)
		Kyoto-Protocol (1997–98)
Danube River Protection Regime 1985–98	1991 and 1994	Danube River Protection (1985–91) (1991–94) (1994–98)
Desertification Regime 1994–98		United Nations Convention to Combat Desertification (1994–98)
Great Lakes Management Regime 1972–98	1978	Great Lakes Water Quality (1972–78) (1978–98)
		Great Lakes Water Quantity (1972–78) (1978–98)
		Great Lakes Ecosystem Management (1978–98)
Hazardous Waste Regime 1989–98	1995	Basel Convention (1989–95) (1995–98)
		Amendment to the Basel Convention (1995–98)
		OECD/EU/Lome IV-Regulations (1989–95) (1995–98)
		Bamako Convention (1991–95)
		Bamako/Waigani Conventions (1995–98)
IATTC Regime (Interamerican Tropical Tuna Convention) 1949–98	1976	Conservation and Management of Tunas and Tuna-Like Fishes (1949–76) (1976–98)
		Conservation and Management of Dolphins (1976–98)
ICCAT Regime (Conservation of Atlantic Tunas) 1966–98		ICCAT-Convention (1966–98)
Regime for the International Regulation of Whaling 1948–98	1982	Whaling Regime (1946–82) (1982–98)

Table 5.1 Continued

Regime	Watershed(s)	Regime Elements
London Convention Regime 1972–98	1991	Wastes and Substances the Dumping of which is Prohibited (1972–91) (1991–98)
		Wastes and Substances which, in Principle, may be Dumped (1972–91) (1991–98)
		Regulation of Incineration at Sea (1978–91) (1991–98)
ECE-Regime on Long-Range Transboundary Air Pollution (LRTAP) 1979–98	1982	LRTAP-Convention (1979–82) (1982–98)
		First Sulphur Protocol (1985–98)
		Nox-Protocol (1988–98)
		VOCs-Protocol (1991–98)
		Second Sulphur-Protocol (1994–98)
North Sea Regime 1972/74–98	1984/1992	OSCOM/PARCOM (1972/74–84)
		OSCOM/PARCOM/OSPAR (1984/92–98)
		North Sea Conferences (1984–98)
Oil Pollution Regime 1954–98	1973/1978	Oilpol (1954–78)
		MARPOL (1973/78–98)
		Regional Memoranda of Understanding (1982–98)
Regime for Protection of the Rhine Against Pollution 1963–98		Berne Convention (1963–98)
		Chloride Pollution Convention (1976–98)
		Chemical Pollution Convention (1976–98)
		Ecosystem/Salmon (Rhine Action Plan) (1987–98)
Ramsar Regime on Wetlands 1971–98	1987	Ramsar Convention (1971–87) (1987–98)
Regime for Protection of the Black Sea 1992–98		Bucharest Convention and Protocols (1992–98)
		Black Sea Strategic Action Plan (1996–98)
South Pacific Fisheries Forum Agency Regime 1979–98	1982 and 1995/97	General Management of Fisheries (1979–82) (1982–95/97) (1995/97–98)
		Compliance of Fisheries Management (1979–82) (1982–95/97) (1995/97–98)
Stratospheric Ozone Regime 1985–98	1990	Vienna Convention (1985–90) (1990–98)
		Montreal Protocol (1987–90) 1990–98)
		London Amendment (1990–98)
		Copenhagen Amendment (1992–98)
		Multilateral Fund (1990–98)
Tropical Timber Trade Regime 1983–98		International Tropical Timber Agreement (1983–98)

the time horizon that applied for the coding of regimes. Together with the regimes on international trade in tropical timber, the conservation of Atlantic tunas, or fisheries in the Barents Sea these regimes consist of only one regime element. However, quite comprehensive case structures emerged for a number of regimes which are based on complex legal arrangements or which have been determined by various legal or political dynamics during their life-cycles. A case in point is the Antarctic regime. This regime not only consists of various components, but it also experienced two watersheds which finally led to the coding of 12 regime elements. While no other regime included in the database involves such a comprehensive case structure, for three regimes pre-coding negotiations led to the identification of seven regime elements. Middle-sized case structures involving between six and four regime elements emerged for eight regimes. Case structures that consist of two or three regime elements have been developed for seven regimes.

Important Actors

States and various types of non-state actors influence the formation or further evolution of regimes. These actors can take roles as pushers or laggards at different stages of the political process. But states and non-state actors can also participate in a regime's programmatic activities. The precoding negotiations were used for identifying important actors. Each of these important actors has been coded in terms of their roles during various stages of the political process. The coding of state behavior focused on a number of topics: the compliance behavior, whether the cognitive understandings about the causes and effects of problems have changed in important states, or whether understandings about the causes and effects of problems or about policy options changed in these states. For some states, data are available for a number of regimes. The United States of America were identified as an important state in 63 of the 92 possible regime elements. Other frequently listed states involve Germany (40), which is followed by the former USSR and the Russian Federation (35), the United Kingdom (35), Japan (23), Australia (22), and Denmark (21). Developing countries like India (16), China (14) or Brazil (10) have also been listed in various regime elements. Altogether 58 important states and the European Union were identified as important actors during these negotiations. The list of important non-state actors or networks of non-state actors is a lot more diverse. It covers over 100 transnational organizations. Some non-state actors were important in a broad number of regimes. Greenpeace has been identified as an important non-state actor in 46 of the 92 regime elements. It is followed by WWF (18), Friends of the Earth (15), or UNEP (12) and IUCN (11). These non-state actors were coded for more than one regime. But the broad majority of non-state actors played an important role in only one of the 23 regimes. The data about non-state actors were complemented by data on nearly 100 individuals which acted as representatives of non-state actors or as members of national delegations or international organizations.

Coding Procedure and Data-Analysis

The 23 cases represent a set of regimes which are managing various types of environmental conflict about collective goods, common pool resources, shared natural resources, or trans-boundary externalities. The geographical scope of the different regimes involves bilateral relations, regional contexts, or multilateral and global politics. Many of these institutions regulate issues that need to harmonize conflicting interests between developed and developing countries. These cases vary with regard to a number of characteristics that become relevant in exploring our dependent and independent variables. Variance exists between these regimes with regard to the degree of effective problem-solving or goal-attainment, compliance behavior, participation of non-state actors, or mechanisms provided for participation by non-state actors. Such variance can also be found within many regimes, if the case structure allows for distinction of different time periods. Changes that affect the understanding about cause-effect relationships, the level of problem-solving, goal-attainment, or compliance behavior can occur within regimes from one time period to another.

The coding has been carried out on the basis of a comprehensive codebook. This book includes variables dealing with the characteristics of regime formation, regime attributes, regime consequences and dynamics (Breitmeier, Levy, Young and Zürn 1996a).[5] Various drafts of this codebook have been tested in trial runs. The trial runs led to the expansion and refinement of rules which have been used for structuring a regime as a case and caused the creation of additional rules that governed the coding process. These trial runs raised a number of methodological issues that are connected to the development of the codebook or the coding process itself. They were determined by the following question: Which measures can be taken that will guarantee the reliability of data provided by coding case study experts? In some respects, the process of developing this codebook has been determined by the insight that in macropolitical research "data are made, not born" (Singer 1990, 9). The generation of data by social scientists is carried out against the background of theoretical concepts or plausible hypotheses which can be operationalized and transformed into measurable variables. Such procedure is confronted with the task to reduce the complexity of theoretical approaches and to translate them into empirically measurable categories. The codebook has been developed with reference to a broad number of theoretical concepts produced by regime analysis. The trial runs could be used to test whether the questions that were raised in the codebook are comprehensible to coding experts. They also served to explore whether operationalizations which translate theoretical concepts into nominal or ordinal scales were plausible.

---

5   For a review of dependent and independent variables that determined the study of international regimes see Levy, Young and Zürn (1995). This review also influenced the development of the IRD-codebook.

One result of the trial runs involved that the coding would result in three types of data. *Nominal scales* were produced for a broad set of variables which refer to the formation and management of regimes. Such nominal scales consist of lists of keywords which were derived from an extensive review of the scientific literature on international institutions. The development of *ordinal scales* was feasible for variables where the analyst intended to distinguish between various degrees of actors' involvement in political processes or between different levels of problem-solving or compliance. The development of nominal and ordinal scales mainly relied on operationalizations which were developed by the research community. But no empirically measurable operationalizations were available for a number of variables (e.g., broader regime consequences). Case study experts were asked to write *narratives* in order to collect more empirical information about these variables. These textual responses can be used by other analysts for constructing nominal or ordinal scales on various dimensions of regime consequences. The combination of various types of data also has some implications for statistical analysis. Direct correlation between variables can not be made if information for one of the variables has been provided in forms of narratives. The data that describe the role of non-state actors during various stages of the political process frequently provide a starting-point for qualitative research. They illustrate the role of a few important non-state actors during regime formation and subsequent stages of the political process.

Discussions which we had with participants in the trial runs led us to conclude that each regime must be coded by two case study experts independently. It can not be excluded that a coding expert will arrive at conclusions concerning the impact of a regime on compliance, problem-solving, or distributional consequences that partly differ from those made by other experts. Accordingly, the availability of two codings allows checking of the data-set for intercoder-reliability. This method enables us to test how a second case study expert considers developments in the issue-area. But the possibility to check the intercoder-reliability of data has also its limits. Coding case study experts were encouraged to refrain from the coding of variables when they felt unable to provide reliable codes. If one of the two coders decided to refrain from the coding of a variable, data analysis could rely on only one coding and had to sacrifice a check for intercoder-reliability. If both coders felt unable to code a variable, no information is included in the database about this variable for a particular regime element. The size of variables which had to be coded was considerable. Differences between the coding of specific variables provided by the two case study experts occurred on various occasions. Many of these differences could be cleared up in discussions with coding case study experts. The coding process passed through several steps which helped to further improve the quality of data, to resolve differences in the codings provided for a regime, and to clear up uncertainties occurring on the part of the coders concerning the meaning of variables, or the operationalizations or scales used for the coding.

Data reliability depends on the competence of coding experts. The research team conducted surveys with some dozens of international scientific, legal, or policy experts with the aim of collecting knowledge about possible cases and coding experts. The selection of case study experts has been determined by the rule of thumb that at least one political scientist or international lawyer should be involved in the coding of a regime. These experts should combine expertise on a regime's legal framework or about political processes that determined the creation and further evolution of regimes. The broad majority of regimes has been coded by scientific experts from the IR-community. These experts have a comprehensive empirical knowledge about the relevant case which they coded and about the theoretical approaches of regime analysis. Policy-experts (e.g., leading officers of regime secretariats, individuals who participated in regime negotiations as members of national delegations) occasionally joined coding IR-scholars and agreed to take roles as coders. The coding of 23 environmental regimes has been a truly interdisciplinary enterprise. The coding revealed that the codebook was comprehensible also to practitioners in international governance systems and to national experts who (whether their professional background was that of an environmental engineer or that of a natural scientist) were not directly involved in academic debates of the IR-community.

A first joint coding round involved the creation of consensus between coding experts about the most important goals, principles, norms and rules, programmatic activities or bodies of a regime. Such agreement guaranteed that the two case study experts shared common inter-subjective understandings about the legal and organizational framework of a regime. Nevertheless, about 90 per cent of the variables included in the database were coded by experts of a regime independently of one another. The decision that each regime was be coded by two case study experts independently had a number of implications for the coding and for data-analysis. If all coders completed their work, the set of regime elements coded for the regime database would have doubled from 92 to 184. Only for 21 of the total of 23 regimes the coding was completed by both coders engaged in this process. Since the coding of the regimes on long-range trans-boundary air pollution in Europe and on fisheries in the South Pacific region was completed only by one coding expert, the regimes database includes data about 172 regime elements.

Chapter 6

Regimes and the Reduction of Uncertainties

The contribution of international institutions to the reduction of uncertainties which prevent that states can manage trans-boundary environmental problems effectively has been referred to as one among a number of grounds which justify obedience. Uncertainties can occur in many different ways. Special attention will be given to uncertainties in the cognitive setting. Cognitive factors can influence the evolution of state preferences and regime policies. Functional regime theory or knowledge-based approaches emphasize the role of institutions for the reduction of transaction costs, for raising concern for trans-boundary problems, for the development of new policies, or for capacity-building (Keohane 1984; Levy, Keohane and Haas 1993). These conceptions of regimes are based on the assumption that preferences of self-interested actors are not fix but can be influenced by a changing knowledge-base or by rising incentives provided by a regime to participate in the management of trans-boundary environmental problems. The programmatic activities of regimes which focus on improving the knowledge about the causes and effects of a problem or about the policy options that are available for problem solving can be understood as social processes "in which scientists, policymakers, and other stakeholders are (or are not) gathering data, conducting analysis, explaining, debating, learning, and interacting which each other around the issue on which the assessment focuses" (Clark, Mitchell and Cash 2006, 14).

In the following, several steps will be taken to clarify whether institutional mechanisms and non-state actors contributed to reduce various dimensions of uncertainty in global governance systems. First, attention will be given to the cognitive setting or to the changes in this setting which determined political processes in environmental regimes. It will be explored how far changes in the understanding about the nature of the problem, in the completeness of information about policy options, or in the capacities of states to participate in social practices on the level of a regime could be observed. Subsequently, the two explanatory approaches will be tested in regard to whether they can explain the changes that could be observed in the cognitive setting. Our puzzle leads us to explore whether changes which can be observed in the cognitive setting correlate with changes in the institutional settings of regimes or with the degree of participation by non-state actors during various stages of the political process. This will inform on whether regimes facilitate the exchange of information and the coordination of monitoring and research. It will also demonstrate whether regimes promote the networking of transnational experts or other groups whose activities aim to get new causal beliefs accepted by the broader political public and by policymakers. From a methodological perspective, special attention will be given to the question about the degree to which agreement exists between the codings provided by two case study experts for a regime element. This methodological discussion will primarily focus on data referring to the dependent variable, since these data have been collected without prior consultation between the two coding experts.

Knowledge, Policy Options, Capacities

Did regimes improve the consensual knowledge about the nature of the problem? Did regimes lead to the development of policy options for political management? Regimes are understood as institutions for the management of one or various problems. These problems were identified together with coding experts in precoding negotiations. For example, such problems involve the concern for the state of the Baltic Sea as a very vulnerable and sensitive area, the increase of greenhouse gases in the atmosphere and inability of humans and ecosystems to adapt to the impacts of climate changes that ensue, or exports and imports of hazardous waste from industrialized to developing countries. The coding that followed the precoding negotiations was carried out to assess each problem for the time period of a regime element under the following questions: How far did the understanding about the nature of the problem develop? How complete was the information about policy options? The 92 regime elements identified for the set of 23 regimes involve a total of 124 problems. One and the same problem has frequently been coded for different regime elements (for example, when a regime component was separated into different time periods). More than one problem existed in some regime elements. Two independent codings normally exist for a regime element. These two different data-sets allow checking of the data for intercoder reliability. If each regime would have been coded by two experts, data about a total of 248 problems would be available. Only 21 of the 23 regimes have been coded by two case study experts independently of one another. Two regimes were coded by only one expert. Accordingly, a total of 235 problems has been coded for 172 regime elements. A double data set exists for 80 of the 92 regime elements. The 235 problems can be distinguished in two different sets of data: two independent codings exist for a set of 111 problems; a set of 13 problems was coded by only one expert. The figure of 235 problems represents a maximum. If case study experts avoided the coding of problems in specific regime elements, the total amount of problems coded can be below this maximum.

In the following, we will distinguish between frequency distributions that can be found either for all cases or for the subset of cases ending in 1998. This distinction will inform how strongly these uncertainties were reduced by the end of the twentieth century. The focus of analysis on cases which end in 1998 makes sense also for methodological reasons. By considering regime elements which end in 1998, a distortion of empirical findings can be avoided. Regimes with one or more watersheds are proportionally over-represented in the total of data, because measurements were made for various time periods. For regimes with no watersheds, only one measurement exists for a regime component. The problem of statistical overrepresentation of data from regimes with watersheds can be avoided by exploring only periods of regimes which end in 1998.

Understanding of the Nature of the Problem

In exploring the understanding of the nature of the problem, different aspects must be considered: the degree of consensus that could be found in the issue-area about the nature, causes and consequences of the problem; the degree of consensus about solutions; or the degree of consensus about the general framing in terms of what should be maximized in the issue-area (e.g., whether members attached great value primarily to the conservation or to the harvesting of a resource). Our first measurement concerning the understanding about the nature of the problem distinguished very strongly or not at all established understandings as the upper and lower limits on an ordinal scale from one to five. It intends to explore how far this understanding about problems emerged in regime elements. Such data exist for 221 problems (see Table 6.1). These problems belong to 168 of our total of 172 regime elements and cover all 23 regimes. For more than two thirds (151) of these 221 problems, a strongly (94) or a very strongly (57) established understanding about the nature of the problem emerged. For 70 problems, this understanding was only partially (47) established or has been low (23).

Two hundred of these 221 data can be checked for intercoder reliability, because 100 pairs of problems were coded by two experts. For 46 of these 100

Table 6.1   Level of Understanding of the Nature of the Problem

All problems		87 pairs of problems (codings at most differing by one value)*		48 pairs of problems which end in 1998 (codings at most differing by one value)*		Level of Understanding of the Nature of the Problem
57	(25.8%)	48	(27.6%)	19	(19.8%)	Very strongly established
94	(42.5%)	80	(46.0%)	63	(65.6%)	Strongly established
47	(21.3%)	35	(20.1%)	11	(11.5%)	Partially established
23	(10.4%)	11	(6.3%)	3	(3.1%)	Low established
0	(0.0%)	0	(0.0%)	0	(0.0%)	Understanding not at all established
221	**100.0%**	**174**	**100.0%**	**96**	**100.0%**	**Total**

* Two codes exist for each problem in a regime element. Pairs of problems were omitted, if disagreement between coding experts has reached more than one value on the scale. Accordingly, the subset of 87 pairs of problems ignores 13 pairs (or 26 data). The subset of 48 pairs applying to regime elements which end in 1998 ignores 7 pairs of problems (or 14 data). Percentage figures were rounded up and down.

pairs of problems identical answers were given by coding experts on a scale from one to five. Forty-one of these 100 pairs differed by one value on this scale. For 13 problems differences were larger than one. This means that findings for a set of 174 data (87 problems each coded by two separate experts) are based on complete agreement or on marginal differences between coding experts. Compared with our total data set, minor differences occur in the frequency distribution for this sub-set of problems. Table 6.1 illustrates that from the subset of 174 data consisting of codings with complete agreement or marginal differences, very strongly (48) or strongly (80) established understandings were indicated for the broad majority of data. Partially or low established understandings were indicated for 46 problems. This suggests that strongly or very strongly developed understandings were identified for more than 73 per cent of the 174 problems.

Which findings emerge for the subset of regime elements ending in 1998? For a subset of 55 pairs of problems ending in 1998 an intercoder reliability check can be made. It reveals complete convergence between both experts in 26 and marginal differences by one value in 22 instances. Seven pairs of answers differ by two values. Table 6.1 illustrates the data of the 48 pairs of problems ending in 1998 where either complete agreement exists or only marginal differences emerged between coding experts. Very strongly (19) and strongly developed understandings (63) make up more than 85 per cent of this total.

Compared to the previously described data set of 174 problems, the data for regime elements ending in 1998 provide an a little less favorable picture as far as the upper limit of our scale is concerned. For only less than 20 per cent of the problems which were coded for regime elements which end in 1998, a very strongly established understanding has been identified. But 27 per cent from the larger set of 174 data indicated a very strongly established understanding. This difference occurs because the majority of regime elements coded for the Antarctic regime ends before 1998. Very strongly or strongly established understandings existed for most of these problems. Nearly two thirds of these pairs of problems were coded for periods which end before 1998. This is caused by the fact that two watersheds were identified in this regime leading to an overrepresentation of cases which end before 1998.

The data which illustrate developments for regime elements ending in 1998 reveal that for the vast majority of regimes strongly or very strongly established understandings emerged. In a number of cases where regime components have been distinguished by watersheds, analysis reveals that these understandings improved in the course of time. The data illustrate that such improvements occurred for cases like the CITES-regime, the Danube regime, the Baltic Sea regime, or for regimes dealing with climate change, the protection of the ozone layer, or whaling. In some regimes or regime components these understandings have reached a high level already from the outset. A case in point is the Antarctic regime. Problems like jurisdictional differences and conflicts about overlapping claims on the part of claimant states and between claimants and non-claimant states were very strongly understood already when the Antarctic Treaty had been concluded in 1959. However, for a few cases like the tropical timber trade regime, the North Sea regime or the hazardous waste regime less progress has been made regarding these understandings.

The above-described data can not fully clarify the dynamics that determined the evolution of this understanding. A second measurement focuses on the changes that occurred for all states as well as on changes that took place in important nations. It distinguishes between significant and little or no change occurring in this understanding during the time period of a regime element. Such change can influence the formation of preferences by domestic societies or governments in a positive way. Data about change that affects understandings about the nature of the problem exist for 204 problems in the database. They belong to 155 regime elements which cover all 23 environmental regimes (see Table 6.2). For 105 problems data indicated a significant change in the understanding of the nature of the problem. Little or no change occurred for understandings of 99 problems. More than 60 per cent of the data which indicate that little or no change occurred apply to the Antarctic regime. For the vast majority of problems very strongly established understandings had existed since the creation of this regime in the late 1950s. Accordingly, the absence of significant change in most regime components results from the fact that these problems were well understood from the beginning.

Seventy-nine pairs of problems were coded by two case study experts. Agreement or minor differences between the data on pairs of problems exist for a total of 138 values (or 69 pairs of problems). Disagreements among codings by more than one value exist for 20 values (or 10 pairs of problems). The answers provided for the subset of 138 values which reflect agreement between coders are equally distributed on the two scales. A significant change in the understanding about the nature of the problem has been diagnosed for 70 problems. Nearly the

Table 6.2   Change in Understanding of the Nature of the Problem (Level of All Regime Members)

All Problems		69 pairs of problems (codings at most differing by one value)*		40 pairs of problems which end in 1998 (codings at most differing by one value)*		Change of Understanding at Level of all Regime Members
99	(48.5%)	68	(49.3%)	32	(40.0%	Little or no change
105	(51.5%)	70	(50.7%)	48	(60.0%)	Significant change
204	**100.0%**	**138**	**100.0%**	**80**	**100.0%**	**Total**

* Two codes (pair of problems) exist for each problem in a regime element. Pairs of problems were omitted, if disagreement between coding experts has reached more than one value on the scale. Accordingly, the subset of 69 pairs of problems ignores 10 pairs (or 20 data). The subset of 40 pairs applying to regime elements which end in 1998 ignores 5 pairs of problems (or 10 data). Percentage figures were rounded up and down.

same amount of answers (68) applies to little or no change. Forty-eight of the data from this subset which indicate a significant change apply to regime elements which end in 1998. The vast majority of data indicating little or no change refer to problems which are managed by various components of the Antarctic Treaty system. The significant changes can be observed for regime elements which belong to 12 regimes. This suggests that for a broad number of regimes the understanding about the nature of the problem has been determined by significant change during periods which ended in the late 1990s.

For cases like the CITES-regime, the ozone regime, the Great Lakes-regime, the London Convention-regime, the Baltic Sea regime or the Danube river protection regime such changes occurred throughout various time periods. For other regimes no change has been detected during more recent periods. But these regimes were affected by changes in the understandings about the nature of the problem in earlier times. For example, the whaling regime has been determined by change in the understanding about the conservation of whale stocks primarily during its earlier period between 1946 and 1982.

Did regimes play a causal role for this change? The causal relevance of a regime for this change has been coded on a scale from one to three. Significant and little causal relevance of a regime were used as upper and lower limits on this ordinal scale. The coding of the causal influence of a regime on various outcomes or impacts demanded of experts complex causal judgments. In making these judgments, coding experts had to weigh regime factors and non-regime factors against each other. Regime factors which stem from a regime's existence may involve that new knowledge has been produced by international scientific collaboration. Non-regime factors operate outside a regime's environment and include factors which are not attributable to the existence of a regime. A total of 203 data describes the causal impact of regimes on the change of understandings about the nature of the problem (see Table 6.3). A moderate or really important causal influence on this change was certified in the predominant number of problems coded. The understanding about the nature of the problem changed significantly for more than

Table 6.3   Causal Impact of Regimes for Changes in the Understandings of the Nature of the Problem (Level of All Regime Members)

| All Problems | Causal Influence of Regimes | | | | Change in Understanding of the Nature of the Problem |
	Not applicable (No change)	Little or no causal impact	Modest causal impact	Significant causal impact	
98 (48.3%)	38 (100.0%)	21 (72.4%)	8 (18.2%)	31 (33.7%)	Little or no change
105 (51.7%)	0 (0.0%)	8 (27.6%)	36 (81.8%)	61 (66.3%)	Significant change
203 100.0%	38 100.0%	29 100.0%	44 100.0%	92 100.0%	Total

Table 6.4   Causal Impact of Regimes for Changes in the Understandings of the Nature of the Problem in Important States

Total Problems	Causal Impact of Regimes								Changes in the Understandings of the Nature of the Problem	Important Nation
	Not applicable (No change)		Little or no causal impact		Moderate causal impact		Significant causal impact			
61	31	(50.8%)	0	(0.0%)	0	(0.0%)	30	(49.2%)	Little or no change	Argentina
10	0	(0.0%)	3	(30.0%)	2	(20.0%)	5	(50.0%)	Significant change	
67	32	(47.8%)	3	(4.5%)	2	(3.0%)	30	(44.7%)	Little or no change	Australia
17	0	(0.0%)	3	(17.6%)	5	(29.4%)	9	(52.9%)	Significant change	
3	1	(33.3%)	2	(66.6%)	0	(0.0%)	0	(0.0%)	Little or no change	Brazil
15	0	(0.0%)	0	(0.0%)	2	(13.3%)	13	(86.7%)	Significant change	
2	0	(0.0%)	0	(0.0%)	2	(100.0%)	0	(0.0%)	Little or no change	Canada
28	0	(0.0%)	2	(7.1%)	4	(14.2%)	22	(78.6%)	Significant change	
6	1	(16.7%)	3	(50.0%)	1	(16.7%)	1	(16.7%)	Little or no change	China
12	0	(0.0%)	2	(16.7%)	5	(41.7%)	5	(41.7%)	Significant change	
9	3	(33.3%)	0	(0.0%)	2	(22.2%)	4	(44.4%)	Little or no change	Denmark
26	0	(0.0%)	0	(0.0%)	16	(61.5%)	10	(38.5%)	Significant change	
8	3	(37.5%)	4	(50.0%)	0	(0.0%)	1	(12.5%)	Little or no change	European Union
18	0	(0.0%)	4	(22.2%)	11	(61.1%)	3	(16.7%)	Significant change	
13	2	(15.4%)	3	(23.1%)	5	(38.5%)	3	(23.1%)	Little or no change	Germany
52	0	(0.0%)	4	(7.7%)	27	(51.9%)	21	(40.4%)	Significant change	
12	2	(16.7%)	7	(58.3%)	1	(8.3%)	2	(16.7%)	Little or no change	France
18	0	(0.0%)	4	(22.2%)	8	(44.4%)	6	(33.3%)	Significant change	
9	1	(11.1%)	6	(66.7%)	1	(11.1%)	1	(11.1%)	Little or no change	India
11	0	(0.0%)	2	(18.2%)	5	(45.5%)	4	(36.4%)	Significant change	
19	0	(0.0%)	12	(63.2%)	6	(31.6%)	1	(5.3%)	Little or no change	Japan
29	0	(0.0%)	3	(10.3%)	17	(58.6%)	9	(31.0%)	Significant change	

Table 6.4   Continued

Total Problems	Not applicable (No change)	Causal Impact of Regimes			Changes in the Understandings of the Nature of the Problem	Important Nation
		Little or no causal impact	Moderate causal impact	Significant causal impact		
58	32 (55.2%)	0 (0.0%)	2 (3.4%)	24 (41.4%)	Little or no change	New Zealand
15	0 (0.0%)	3 (20.0%)	3 (20.0%)	9 (60.0%)	Significant change	
18	3 (16.7%)	10 (55.6%)	2 (11.1%)	3 (16.7%)	Little or no change	Norway
7	0 (0.0%)	0 (0.0%)	1 (14.3%)	6 (85.7%)	Significant change	
3	0 (0.0%)	2 (66.7%)	1 (33.3%)	0 (0.0%)	Little or no change	Poland
18	0 (0.0%)	0 (0.0%)	6 (33.3%)	12 (66.7%)	Significant change	
7	3 (42.9%)	2 (28.6%)	2 (28.6%)	0 (0.0%)	Little or no change	Sweden
17	0 (0.0%)	0 (0.0%)	5 (29.4%)	12 (70.6%)	Significant change	
10	2 (20.0%)	3 (30.0%)	5 (50.0%)	0 (0.0%)	Little or no change	Switzerland
14	0 (0.0%)	3 (21.4%)	7 (50.0%)	4 (28.6%)	Significant change	
74	35 (49.3%)	5 (6.8%)	1 (1.4%)	33 (44.6%)	Little or no change	United Kingdom
31	0 (0.0%)	6 (19.4%)	9 (29.0%)	16 (51.6%)	Significant change	
94	38 (40.4%)	12 (12.8%)	10 (10.6%)	34 (36.2%)	Little or no change	United States
67	1 (1.5%)	12 (17.9%)	22 (32.8%)	32 (47.8%)	Significant change	
68	32 (47.1%)	2 (2.9%)	4 (5.9%)	30 (44.1%)	Little or no change	USSR/Russian Federation
46	0 (0.0%)	6 (13.0%)	14 (30.4%)	26 (56.5%)	Significant change	

half of the problems. In nearly 60 per cent of the instances where a particularly strong change has been measured the regime performed a significant causal influence. On balance, the causal influence detected for a regime for significant change in these understandings seems to lie between modest and significant.

Other findings which describe the changing understanding of the nature of the problem in important states support the conclusion that regimes had an impact in many cases (see Table 6.4). Between four and seven important states identified during the precoding negotiations were coded for each regime element. One of the questions in the codebook asked whether important states were affected by such change. For example, countries like the United States, the United Kingdom, Germany, Canada, Australia, Argentina, Denmark or Russia were coded for a dozen or several dozens of regime elements. One hundred and sixty-one data describe such changes in the United States. The vast majority of the 94 data which indicate that little or no change in the United States refers to problems that were coded for the Antarctic regime. Exploring changes for the United States on the basis of a subset of 15 regimes which ignores the Antarctic regime leads us to a different conclusion. For this subset, little or no change (32) has been identified less frequently than significant change (57). Measurements made for Russia (or the former Soviet Union) and for Germany lead to findings that are similar to findings identified for the United States. For Russia or the Soviet Union as its predecessor, 114 problems were coded. The majority of data indicates that the understanding in Russia did not change. But the picture changes when only those five regimes other than the Antarctic are considered. From a total of 42 data existing for this set of five regimes, only 6 indicate little or no change and 36 of them identify significant change. The vast majority of answers indicates that a modest or significant causal influence existed for a significant change that determined Russia's understanding about the nature of the problem. The fall of the iron curtain and the collapse of communism made the political system more open-minded for information about environmentally damaging practices revealed by the activities of international institutions or non-state actors. It became more easier for transnational or domestic environmental actors to verify alleged pollution of the environment such like dumping of high- or low-level radioactive waste at seas (Meinke 2002, 205–7).[1]

Significant change has obviously determined Germany's understanding of the nature of various trans-boundary environmental problems. For 52 of the 65 problems coded a significant change in Germany's understanding has been identified. This means that a change took place in the predominant number of the nine regimes that were coded for Germany. In 48 of the 52 instances with a

---

1   It has been argued in a study about nuclear dumping in arctic seas that "Soviet and Russian management of nuclear waste in the north has been significantly influenced by regulations and programs generated under the London Convention" and that "as a result of the regime, scientific surveys and monitoring of the radiological situation around the dumping sites have been stimulated" (Stokke 1998a, 476).

significant change the regime had a modest (27) or significant (21) causal influence. Similar findings exist for countries like Canada, Japan, Poland or Sweden. The majority of data indicating little or no change which occurred for countries like the United Kingdom, Australia or New Zealand is caused by overrepresentation of problems coded for the Antarctic regime where understandings had been strongly developed from the outset. For a group of countries consisting of China, India or Brazil a significant change in the understanding has been identified in 38 instances. In the broad number of cases, regimes had a modest (12) or significant (22) causal impact. Little or no change occurred in 18 instances.

Policy Options

The development of policies in international regimes is based on information about options which can be chosen for the management of trans-boundary problems. Such options can focus on measures which must be taken in different sectors. Information about such policy options must normally have reached a special level of completeness which convinces policymakers that they are implementing the most effective measures, that proposed solutions can be implemented without provoking serious resistance by interest-groups, or that these policies will lead to desired consequences. How complete was information about these policy options? How far was this information determined by change? Did the regime play a causal role for this change? An ordinal scale from one to three has been used for the coding of different levels of the completeness of the quality of this information. Very high or low completeness were distinguished as upper and lower limits on this scale.

Table 6.5   Completeness of Information about Policy Options

All problems		77 pairs of problems (codings at most differing by one value)*		46 pairs of problems which end in 1998 (codings at most differing by one value)*		Completeness of Information about Policy Options
37	17.9%	28	18.2%	20	21.7%	Very high completeness
133	64.2%	104	67.5%	65	70.7%	Medium completeness
37	17.9%	22	14.3%	7	7.6%	Low completeness
207	**100.0%**	**154**	**100.0%**	**92**	**100.0%**	**Total**

* Two codes exist for each problem in a regime element. Pairs of problems were omitted, if disagreement between coding experts has reached more than one value on the scale. Accordingly, the subset of 77 pairs of problems ignores 5 pairs (or 10 data). The subset of 46 pairs applying to regime elements which end in 1998 ignores 4 pairs of problems (or 8 data). Percentage figures were rounded up and down.

The degree of completeness of information about policy options was measured for 207 problems. They belong to 166 regime elements. All 23 regimes are covered by these regime elements. From the total of 207 data, 37 indicated a very high completeness of information about policy options. Medium levels of completeness had been achieved in 133 and low levels in 37 instances (see Table 6.5). From the subset of 82 problems coded by two experts, complete agreement exists between 39 pairs of problems. For another set of 38 pairs of problems, codings vary by one value. Five pairs of data were eliminated because differences make up two values on our scale.

If we ignore those data where only one data set exists or where differences within pairs of problems are larger than one, 28 represent very high and 104 medium completeness, whereas 22 have been coded for low completeness. At first sight, these findings suggest that information about policy options has advanced to a lesser extent than we may wish in light of the far-reaching measures which are required to manage environmental problems effectively. For nearly half of the cases included in the subset of 46 pairs of problems that were coded for periods which end in 1998, both experts agreed that only a medium degree of completeness existed. Such medium levels of completeness existed by the end of the twentieth century, for example, for several components of the hazardous waste regime, the Black Sea regime, the North Sea regime, the London Convention regime, the climate change regime, or the Baltic Sea regime.

Did the completeness of information about policy options change? A second measurement focusing on these changes gives a more dynamic picture (see Table 6.6). Information about such changes exists for 194 problems. The data illustrate that the completeness of information changed significantly in more than half of the instances. 71 pairs of problems (involving a total of 142) were coded by two experts. In comparison with data of other variables less agreement exists among the codings

Table 6.6   Change in Completeness of Information about Policy Options of the Problem (Level of All Regime Members)

Change for All Problems		51 pairs of problems (codings at most differing by one value)*		28 pairs of problems which end in 1998 (codings at most differing by one value)*		Change in Completeness of Information about Policy Options
95	49.0%	60	58.8%	24	42.9%	Little or no change
99	51.0%	42	41.2%	32	57.1%	Significant change
194	**100.0%**	**102**	**100.0%**	**56**	**100.0%**	Total

* Two codes (pair of problems) exist for each problem in a regime element. Pairs of problems were omitted, if disagreement between coding experts has reached more than one value on the scale. Accordingly, the subset of 51 pairs of problems ignores 20 pairs (or 40 data). The subset of 28 pairs applying to regime elements which end in 1998 ignores 9 pairs of problems (or 18 data). Percentage figures were rounded up and down.

Table 6.7   **Causal Impact of Regimes for Changes in the Completeness of Information about Policy Options (Level of All Regime Members)**

| All Problems | Causal Influence of Regimes | | | | Change in Understanding of the Nature of the Problem |
	Not applicable (No change)	Little or no causal impact	Modest causal impact	Significant causal impact	
95 (49.0%)	38 (100.0%)	14 (66.7%)	15 (30.6%)	28 (32.6%)	Little or no change
99 (51.0%)	0 (0.0%)	7 (33.3%)	34 (69.4%)	58 (67.4%)	Significant change
194 100.0%	38 100.0%	21 100.0%	49 100.0%	86 100.0%	Total

of these pairs. Agreement between both experts exists for 51 pairs of problems. For 21 of these pairs (or 42 data) significant changes have been identified. 30 pairs of answers (60 data) applied to little or no change The data set which is characterized by agreement between coding experts indicates that changes in the information about policy options affected a broad number of regimes: the London Convention regime, river regimes for the protection of the Danube or the Rhine, and resource and nature conservation regimes like the Ramsar regime on wetlands, the ICCAT-regime, the tropical timber trade regime, or the Barents Sea fisheries regime.

The data about the causal influence of a regime resemble the results which arose for the impact of regimes on changing understandings of the nature of problems. The regime impact was significant in 58 of the 99 instances where a significant change in the information about policy options was identified (see Table 6.7). A large number of problems have not been affected by such change. Others which were affected significantly by change experienced a moderate or little causal impact by regimes. The findings support the conclusion that information about policy options changed in a broad majority of regimes covered by the IRD. But information has in most of these regimes been less than complete.

Similar changes emerged on the level of important states (see Table 6.8). Data for countries like the United States, Argentina, Australia or the United Kingdom predominantly indicate that no such change concerning the completeness of information about policy options occurred in regard to problems managed by the Antarctic regime. But information about policy options frequently changed in these countries in the context of other regimes. Sixty-nine of the 159 data that exist for the United States indicate that the United States was affected by significant change in seven other regimes. In 56 of the 69 instances, the regime had a modest or significant causal influence on these changes. In some countries (e.g., Canada, Germany, France, Japan, Poland, Sweden, or Switzerland) significant change concerning the information about policy options was the dominant pattern. It occurred more frequently than the pattern "little or no change". Regimes had at

Table 6.8   Causal Impact of Regimes for Changes in the Completeness of Information about Policy Options in Important States

Total Problems		Causal Impact of Regimes			Changes in the Completeness of Information about Policy Options	Important Nation
	Not applicable (No change)	Little or no causal impact	Moderate causal impact	Significant causal impact		
61	35 (57.4%)	0 (0.0%)	0 (0.0%)	26 (42.6%)	Little or no change	Argentina
11	0 (0.0%)	1 (9.1%)	0 (0.0%)	10 (90.9%)	Significant change	
64	35 (54.7%)	2 (3.1%)	1 (1.6%)	26 (40.6%)	Little or no change	Australia
20	0 (0.0%)	2 (10.0%)	4 (20.0%)	14 (70.0%)	Significant change	
13	0 (0.0%)	2 (15.4%)	11 (84.6%)	0 (0.0%)	Little or no change	Brazil
7	0 (0.0%)	1 (14.3%)	2 (28.6%)	4 (57.1%)	Significant change	
13	0 (0.0%)	3 (23.1%)	10 (76.9%)	0 (0.0%)	Little or no change	Canada
16	0 (0.0%)	2 (12.5%)	3 (18.8%)	11 (68.8%)	Significant change	
4	0 (0.0%)	1 (25.0%)	2 (50.0%)	1 (25.0%)	Little or no change	China
12	0 (0.0%)	2 (16.7%)	4 (33.3%)	6 (50.0%)	Significant change	
18	0 (0.0%)	3 (16.7%)	12 (66.7%)	3 (16.7%)	Little or no change	Denmark
14	0 (0.0%)	0 (0.0%)	7 (50.0%)	7 (50.0%)	Significant change	
5	0 (0.0%)	3 (60.0%)	1 (20.0%)	1 (20.0%)	Little or no change	European Union
15	0 (0.0%)	1 (6.7%)	2 (13.3%)	12 (80.0%)	Significant change	
22	0 (0.0%)	8 (36.4%)	11 (50.0%)	3 (13.6%)	Little or no change	Germany
39	0 (0.0%)	3 (7.7%)	18 (46.2%)	18 (46.2%)	Significant change	
7	0 (0.0%)	2 (28.6%)	4 (57.1%)	1 (14.3%)	Little or no change	France
21	0 (0.0%)	3 (14.3%)	3 (14.3%)	15 (71.4%)	Significant change	
6	0 (0.0%)	4 (66.7%)	1 (16.7%)	1 (16.7%)	Little or no change	India
12	0 (0.0%)	2 (16.7%)	4 (33.3%)	6 (50.0%)	Significant change	

Table 6.8   Continued

Total Problems	Not applicable (No change)		Causal Impact of Regimes					Changes in the Completeness of Information about Policy Options	Important Nation
			Little or no causal impact		Moderate causal impact		Significant causal impact		
12	0	(0.0%)	8	(66.7%)	3	(25.0%)	1	(8.3%)	Little or no change	Japan
32	0	(0.0%)	5	(15.6%)	11	(34.4%)	16	(50.0%)	Significant change	
58	35	(60.3%)	0	(0.0%)	2	(3.4%)	21	(36.2%)	Little or no change	New Zealand
15	0	(0.0%)	1	(6.7%)	2	(13.3%)	12	(80.0%)	Significant change	
12	1	(8.3%)	4	(33.3%)	3	(25.0%)	4	(33.3%)	Little or no change	Norway
13	0	(0.0%)	0	(0.0%)	8	(61.5%)	5	(38.5%)	Significant change	
2	0	(0.0%)	2	(100.0%)	0	(0.0%)	0	(0.0%)	Little or no change	Poland
14	0	(0.0%)	0	(0.0%)	12	(85.7%)	2	(14.3%)	Significant change	
3	1	(33.3%)	2	(66.7%)	0	(0.0%)	0	(0.0%)	Little or no change	Sweden
16	0	(0.0%)	0	(0.0%)	14	(87.5%)	2	(12.5%)	Significant change	
3	0	(0.0%)	1	(33.3%)	1	(33.3%)	1	(33.3%)	Little or no change	Switzerland
19	0	(0.0%)	0	(0.0%)	9	(47.4%)	10	(52.6%)	Significant change	
68	36	(52.9%)	6	(8.8%)	0	(0.0%)	26	(38.2%)	Little or no change	United Kingdom
33	0	(0.0%)	8	(24.2%)	10	(30.3%)	15	(45.5%)	Significant change	
90	38	(42.2%)	9	(10.0%)	15	(16.7%)	28	(31.1%)	Little or no change	United States
69	1	(1.4%)	12	(17.4%)	20	(29.0%)	36	(52.2%)	Significant change	
73	35	(47.9%)	0	(0.0%)	12	(16.4%)	26	(35.6%)	Little or no change	USSR/Russian Federation
34	0	(0.0%)	4	(11.8%)	9	(26.5%)	21	(61.8%)	Significant change	

least a modest influence for significant changes. But in many instances, the regime impact was significant. Obviously, different sources that exist within and beyond the nation-state contribute to develop information about policy options. Many countries have comprehensive research capacities. But these capacities alone are often not sufficient to develop new ideas, solutions, or policy options.

Capacities

The evolution of capacities which enable members to participate in social practices at the regime level have been explored on the basis of narratives written by coding experts. Information about a regime's contribution to capacities at the level of member states has been provided for a regime as a whole and often neglects the division into separate regime elements. Instructions were available in the codebook which allowed the determination of possible increases in the capacities of member states to participate in social practices. No such impacts have been identified for five regimes. For some regimes, coding experts disagreed about whether such impacts really exist. The following assessment disregards the set of regimes where such disagreements emerged. No disagreement about the existence of such impacts emerged for 11 regimes. The narratives of coding experts are used to distinguish two particular functions of environmental regimes for raising capacities of members.

Pull for Cooperation and Socialization The institutional framework provided by a regime helped states to promote and further deepen cooperation in the issue-area or to link developments in the issue-area more with other issues. Several narratives support the conclusion that regimes have an important function for enhancing communication between states in the issue area. Institutions provide frameworks for regular exchange among various types of actors. They enable actors at various levels to learn more about the issue and to reduce uncertainties related to various issues during the formation or implementation of regimes. For the London Convention regime, it has been indicated that the "establishment of regular, formal and informal consultations among parties and with intergovernmental and international non-governmental organizations built up over years, and constituting a shared experience on the interpretation and harmonized implementation of the regime is an essential complement to the regime itself".[2] Likewise, a similar role has been ascribed to the Fisheries Forum Agency (FFA) as "an agent for promoting access by the smaller FFA members into the wider global aspects of fisheries and related developments". Since the regime which manages fisheries in the South Pacific had FFA as an international organization at its core, it enabled many member states to participate through FFA in "wider relevant activities such as those of the FAO that would have been beyond their resources, at least on a routine

2 Written narrative on change in the capacities of regime members made by Rene Coenen.

basis".[3] Further on, regimes occasionally contributed that states increasingly understood the functioning of international institutions or were socialized in cooperative frameworks at the international level. For example, the Danube river protection regime gave an opportunity to states which newly emerged after the dissolution of the Soviet Empire in 1989/91 to gain experience on the functioning of international institutions and to learn how to work and behave in relations with the European Union.

Capacity Building for Implementation and Compliance The regime's contribution for improving compliance or for supporting implementation on the domestic level has been referred to as an important achievement. The organization of training sessions which were sponsored by regime bodies or other actors played an important role in the CITES-regime to provide administrators with knowledge and expertise on matters related to trade in endangered species. Likewise, such training courses or the transfer of knowledge and technological assistance were referred to as important for improving implementation of the Danube Convention. In addition, the Memorandums of Understanding which were added to the MARPOL regulations "helped states improve their port state control procedures both by standardizing them and helping states improve their ability to conduct inspections of ships for compliance with relevant treaties".[4] Participation by developing countries in negotiations could be improved in single cases by establishing a trust fund which financed participation of delegations by these countries. Nevertheless, participation by developing countries in scientific and technical meetings can still be constrained by lack of financial resources. Voluntary contributions made by other contracting parties frequently helped to make participation by these countries possible. Such efforts for capacity-building have also been made in the ECE-regime on long-range transboundary air pollution. The demand for these contributions of regimes to raise the capacities of members for implementation and compliance or for participation in institutions varies between institutional contexts. This demand occurs primarily in a context determined by strong asymmetries of financial or other capacities. These asymmetries can be found quite frequently in the context of North-South relations or with regard to participation by, and implementation of policies in, former socialist countries of Eastern Europe. They are irrelevant in relations among industrialized countries.

Summary: Cognitive Uncertainties on the Level of a Regime

By and large, our findings support the conclusion that environmental regimes contributed to improve our knowledge about various aspects in the past few

---

3   Written narrative on change in the capacities of regime members made by Richard A. Herr.

4   Written narrative on change in the capacities of regime members made by Ronald B. Mitchell.

decades. Empirical findings reveal that for a majority of regimes the understanding about the nature of the problem has been strongly or very strongly established at the end of the twentieth century. For the vast majority of regimes where regime elements which end in 1998 could be compared with cases which end no later than 1992, an increase in this understanding could be observed. Many regimes were exposed to significant change leading to improvements that could be reached in this understanding. It could be shown that these changes affected many important states in environmental issue-areas. Information about policy options has mainly reached a medium level of completeness. For more than half of the regimes where comparison between various time periods of a regime is possible, an increase in the completeness of information occurred. Obviously, the cognitive setting of regimes has been determined by strong change during the past few decades. These changes involved that the understanding about the nature of the problem as well as the information about policy options improved on the level of regime members. The causal role ascribed to regimes for change in the cognitive setting is mainly established between modest or significant. Information gathered from written narratives could illustrate that a number of regimes contributed to improve the capacities of members states.

Regimes, Programmatic Activities and the Cognitive Setting

Which contributions of institutional mechanisms influenced the understandings about the nature of the problem? Were there activities through which institutional mechanisms contributed to improve information about policy options? Did these impacts mainly result from the activities of regimes? Or were regimes also dependent on the support of states or non-state actors? Institutional mechanisms can consist of programs which coordinate or expand environmental monitoring and scientific research, which review the adequacy of commitments, or which assess difficulties occurring in connection with the implementation of regime policies. The causal impact of regimes on changes in the different dimensions of the consensual knowledge (understanding about the nature of the problem, information about policy options) has been illustrated by the findings included in Tables 6.3 and 6.7. These findings revealed that regimes played a causal role on the regime level and on the level of important states. In the following, it will be explored whether programmatic activities emerged in regimes that could influence the evolution of consensual knowledge. Such programmatic activities focus, for example, on the scientific monitoring and research of the causes and effects of a problem, on the review of implementation, on reviewing the adequacy of commitments, and on the verification of compliance or compliance monitoring.

Those results which establish a connection between the existence of programmatic activities and the knowledge about cause-effect relationships or about the availability of politics options are contained in Tables 6.9 and 6.10. Table 6.9 correlates data about different levels of understanding about the nature

Table 6.9   Programmatic Activities and Understanding about the Nature of the Problem

Level of Understanding of the Nature of the Problem	Total of Problems	Programmatic Activities				At least one of the four programs existed
		Scientific Monitoring of Causes and Effects	Research about Causes and Effects	Review of Implementation	Reviewing Adequacy of Commitments	
Very strongly established	58 (28.3%)	23 (17.3%)	15 (13.4%)	13 (13.4%)	23 (24.2 %)	40 (69.0%)
Strongly established	82 (40.0%)	61 (45.9%)	54 (48.2%)	49 (50.5%)	40 (42.1%)	67 (81.8%)
Partially established	41 (20.0%)	33 (24.8%)	34 (30.4%)	28 (28.9%)	25 (26.3%)	37 (90.2%)
Low established	24 (11.7%)	16 (12.0%)	9 (8.0%)	7 (7.2%)	7 (7.4%)	20 (83.3%)
Not at all established	0 (0.0%)	0 (0.0%)	0 (0.0%)	0 (0.0%)	0 (0.0%)	0 (0.0%)
Total	205 (100.0%)	133 100.0%	112 100.0%	97 100.0%	95 100.0%	164 (80.0%)

Table 6.10 Programmatic Activities and the Completeness of Information about Policy Options

	Programmatic Activities						
Total of Problems	**Scientific Monitoring of Causes and Effects**	**Research about Causes and Effects**	**Expert Advice**	**Compliance Monitoring**	**Review of Implementation**	**Information about Policy Options**
35 (19.4%)	26 (20.3%)	20 (18.7%)	24 (16.2%)	18 (18.4%)	18 (19.8%)	Very high Completeness
109 (60.6%)	77 (60.2%)	67 (62.6%)	99 (66.9%)	62 (63.3%)	55 (60.4%)	Medium Completeness
36 (20.0%)	25 (19.5%)	20 (18.7%)	25 (16.9%)	18 (18.4%)	18 (19.8%)	Low Completeness
180 (100%)	128 100.0%	107 100.0%	148 100.0%	98 100.1%	91 100.0%	

	Programmatic Activities						
Total of Problems	**Verification of Compliance**	**Financial and Technology Transfer**	**Reviewing Adequacy of Commitments**	**Information Management**	**At least one of the nine programs existed**	**Information about Policy Options**
	22 (22.2%)	7 (14.0%)	15 (16.3%)	14 (14.1%)	34 (97.1%)	Very high Completeness
	56 (58.9%)	33 (66.0%)	62 (67.4%)	69 (69.7%)	108 (99.1%)	Medium Completeness
	17 (17.9%)	10 (20.0%)	15 (16.3%)	16 (16.2%)	34 (94.4%)	Low Completeness
180 (100%)	95 100.0%	50 100.0%	92 100.0%	99 100.0%	176 (97.8%)	Total

of the problem with the existence of programmatic activities. It gets obvious from Table 6.9 that in two thirds of the instances the knowledge about causes and effects or about the real character of the environmental problem has been strongly or very strongly established. A broad number of programmatic activities existed to collect consensual knowledge about the set of 140 problems where strong or very strong understandings emerged. The results existing for this subset show that such programs were set up either for a little more than half (monitoring) or for a little less than half (research, review of the implementation or of the adequacy of commitments) of these problems. It gets also obvious from the data that the variances appearing on the part of the dependent variable are not reflected in the same measure into variances in the independent variable, though. Programmatic activities aiming at improving consensual knowledge existed also for such problems for which the knowledge was only partially developed. In almost all those cases with weakly developed understandings about the nature of the problem one of the four programmatic activities existed at least. Similar problems of explanation arise from the findings in Table 6.10. The knowledge about policy options was established predominantly on a medium level. There was for almost every problem at least one programmatic activity which contributed to improve the knowledge about policy options. The results contained in Table 6.10 show that variances in the quality of information about policy options are not reflected by corresponding variances in the independent variable. Programmatic activities existed for those problems where the information about policy options was low. Obviously, regimes can have an impact. But we need to explain why this impact did not emerge in all cases equally.

Findings included in the two tables illustrate that a broad number of different programmatic activities has been established in environmental regimes. Institutions provided the framework within which scientific and policy-relevant information could be exchanged and resources be pooled. They allowed the development of mid- and long-term research agendas, the improvement of scientific research, or the coordination of monitoring. The broad majority of regimes established programs which contributed to the improvement of the knowledge-base. The data included in the IRD reveal that 20 of the 23 regimes established mechanisms for scientific monitoring of causes and effects. Programs for the scientific research of cause-effect-relationships have been established by 20 regimes. 21 regimes rely on mechanisms through which they can benefit from the advice of experts. Eighteen regimes include each of these three programmatic activities. It should be noted that these programs could not always be found in each element of a single regime. Another impact of regimes on the reduction of uncertainties obviously occurred in connection with the operation of programmatic activities designed for reviewing the adequacy of commitments. Such institutional mechanisms have emerged in 17 regimes. Reviews carried out on a regular basis contribute to expand the debate about the feasibility of existing policies on national and international levels. Each of these regimes also established mechanisms for the review of implementation. Some of these regimes established these reviews only for single regime elements. In a few regimes, the character of the problems did not require the establishment of comprehensive programs for monitoring or scientific research. A case in point is the regime for the prevention of intentional oil pollution at sea. Regime elements involving OILPOL and MARPOL or regional memorandums of understanding were developed to guarantee that regulations aiming at the reduction of oil pollution at sea will be complied with and that compliance behavior can be verified. This regime primarily focused on the creation of mechanisms for reporting, verification, or compliance monitoring (Mitchell 1994). Less attention has been paid by this regime to the creation of new knowledge about the causes and effects of the problem. This function has been continued in coastal states which are affected by oil pollution from tankers.

These programs intend to improve the consensual knowledge. But their existence does not automatically lead to the fulfillment of these goals. This phenomenon can possibly be explained by additional factors. Ecosystems whose protection or sustainable management is intended by international institutions differ with respect to their natural complexity. There are also differences between issue-areas concerning the social contexts or other factors which are responsible for causation and relevant for problem solving. It must also be taken into account that the evolution of the consensual knowledge needs time. In some cases, the production of this knowledge needs more time and efforts than in others. It is possible to compare the level of this knowledge that has been achieved in single regime elements in earlier and later stages of life cycles. These findings reveal that the consensual knowledge improved during the later phases. For example, such developments occurred in the CITES-regime, the Baltic Sea regime or the Danube river regime. In some regimes, the consensual knowledge improved without the existence of explicit programmatic activities. This has mostly been caused by external activities of international institutions or national facilities which executed these functions in charge of regimes.

The broadening of consensual knowledge was an important goal of most regimes. Many regimes included legal provisions which defined collection and dissemination of data, improvement of scientific knowledge, or development of international policies as common goals shared by members. Intensification of scientific monitoring and research was frequently considered as a requirement by regime members for improving consensual knowledge. One of a number of basic goals included in the 1979 ECE-Convention on Long-Range Trans-boundary Air Pollution in Europe involved that "parties shall by means of exchanges of information, consultation, research and monitoring, develop (...) policies and strategies which should serve as a means of combating the discharge of air pollutants".[5] In the Great Lakes management regime the United States and Canada agreed on the creation of bi-national institutional mechanisms to monitor water quality and to promote scientific understanding of the Great Lakes environment, to establish common water quality objectives, compatible standards, or commitments to implementation and procedures for monitoring programs.

---

5   See Art. 3 of the 1979 Convention on Long-range Trans-boundary Air Pollution.

The data included in the IRD illustrate the emergence of programmatic activities. They do not inform about the scope and intensity of their work. A closer look on the development of these programs in single regimes also reveals that the range of activities carried out in single programs broadened over time. The expansion of their work on a broader range of causes and effects frequently coincided with the broadening of a regime's functional scope when it began to address regulatory targets (e.g., pollutants or other environmentally-damaging practices) that had not been considered during the formation of a regime. While scientific monitoring or research normally preceded the creation of new regime components, they remained important also during the implementation phase. The expansion of the range of monitoring and of research activities was necessary for assessing the effectiveness of policies or for identifying new areas where political action was required. The Inter-American Tropical Tuna Commission (IATTC) established two programs through which member states improved their knowledge about relevant aspects related to fishing of tropical tunas and tuna-like species in the Eastern Pacific Ocean (Peterson and Bayliff 1985). The Tuna-Billfish Program has been established to study the biology of tunas and tuna-like-species in this area and to estimate the effects arising from fishing and natural factors on their abundance. These studies led experts to arrive at recommendations about appropriate conservation measures. The Tuna-Dolphin Program was established as a response to growing depletion of dolphins through purse seine fishing where dolphins get injured or killed in the nets used for the fishing of tuna. This program has been established to monitor the abundance of dolphins and their mortality. It involves collection of data aboard tuna purse seiners, analysis of these data or development of recommendations for the conservation of dolphins, and focuses on studying the effects of different modes of fishing on the various fish and other animals of the pelagic ecosystem (IATTC 1999, 73f). Similar comprehensive monitoring and research programs have been established by the vast majority of the regimes explored. In most of the cases, they were closely linked with efforts for the development of policies which can be used for improving the state of environmental problems. In some regimes, these activities have been included under comprehensive frameworks that coordinate the evolution of scientific knowledge, the development of policies, or contribute to national implementation. The Baltic Sea Joint Comprehensive Environmental Action Programme (JCP) established by the members of the Baltic Sea regime in 1992 is a 20-year program of action focusing on the "identification of pollution sources within the entire catchment area of the Baltic Sea and implementation of measures for decisive reduction of emissions and discharges of nutrients and other harmful substances" (HELCOM 1999, 1).

Single states often take roles as lead countries in the implementation of programs which focus on scientific research or on the development of policies. Programmatic activities of a regime closely collaborate with national research facilities and transnational networks. Governments frequently finance additional research activities which contribute to improve the consensual knowledge. Programmatic activities can not be considered as systems which are autonomous of states. States link their own research activities partly to these programs or intend to influence research agendas so that they reflect their own priorities in the issue area. The programmatic activities established in regimes improved access to national data and led to the development of common standards for the collection of data which are relevant for measuring the causes and consequences of environmental problems or for assessing the effectiveness of national policies for implementation. National reports which have to be submitted to regime secretariats on a regular basis are one among a number of steps which can elucidate whether international policies will be implemented and whether problems occur in this process that require the adjustment of international policies to changing socio-economic practices. For example, members of the CITES-Convention emphasized at their 11th meeting of the Conference of the Parties the "importance of annual reports as the only available means of monitoring the implementation of the Convention and the level of international trade in specimens of species included in the Appendices".[6] The guidelines for the preparation of national reports have frequently been standardized so that they allow systematic collection of information required for assessing the implementation in member states. However, one has to admit that neither these guidelines nor deadlines for submission were sufficiently complied with in some regimes.

Insufficient data reporting is a problem which occurs in many regimes. For example, the ozone regime has been affected by insufficient reporting from developing countries during the 1990s. Regime secretariats offer assistance to countries which lack capacities required for the preparation of these reports. But there are frequently problems with the on-time submission of reports. The collection of scientific data or the production of reports about the causes and effects of environmental problems or about the implementation of international policies on the domestic level can not be achieved without the support of national administrations or governments. While information which is required for reviewing national implementation largely refers to practices of private actors, only the state can normally collect data about production methods or emissions caused by industry, about consumption patterns or other information which is relevant for the development of reliable assessments.

The programmatic activities for monitoring, research, or for the review of the adequacy of commitments structure knowledge on the basis of commonly agreed methodologies. Only such commonly accepted standards made it possible to raise the quality of data or to achieve a comparable set of national data on the basis of which the evolution of environmental problems could be assessed, policies for environmental problem-solving developed or adjustments of these policies to newly arising environmental threats be made. Regimes frequently initiated or intensified data collection by states. They brought states to learn more about those causes and effects which could be found in other states or within their own national boundaries. The committees or workshops which take place in regimes

---

6   See CITES-Resolution Conf. 11.17 (rev. COP 12): "Annual Reports and Monitoring of Trade", as Amended at the 12th Meeting of the Parties, Santiago (Chile), 3–15 November 2002.

provide a framework for the discussion of new scientific facts or policy options. Facts do not speak for themselves, but they require interpretation by scientists, technical experts or policymakers.

The Impact of Non-State Actors on the Cognitive Setting

In general, the role of states has remained constant throughout various time periods in regimes. Data do not support the thesis about a possible declining role of the state in international environmental governance systems. The exploration of factors which were most influential for agenda-setting reveals that the impact of states on this process remained constant – whether cases ended before 1992 or only in 1998. Non state actors were present as a factor during agenda-setting in nearly each regime element. In addition, state influence on agenda-setting has been identified as a factor which was most influential for more than half of these regime elements. No other factor has been identified as frequently as state influence. State influence remained the dominant pattern. But various types of non-state actors played important roles during these processes. A similar finding emerges from data which assess whether agenda-setting, negotiations, or the operationalization of agreements where determined by a single state or state coalition, by an inter-state process, or by transnational forces. States have never been removed from their dominant role. But transnational forces could partly increase their influence. Data reveal that the dominance of state influence has been more pronounced during negotiations than during agenda-setting. Growing relevance of non-state actors does normally not occur at the expense of state influence. Non-state actors frequently complement state activities. They also contribute to carry out those functions which could not effectively be managed by states alone. While these data illustrate the role of non-state actors during agenda-setting or negotiations, they raise the question on whether a significant impact of these actors on the production of consensual knowledge during the formation or management of regimes could be observed. The focus of attention will be directed on the activities of scientific or other service organizations. It will also be explored which role national and international activist groups played for communicating this knowledge to the transnational public. Analysis is also directed on economic actors. This can inform on whether national or transnational industrial organizations or corporations became involved in the development of policy options.

Our data reveal that participation by international organizations or national and international scientific organizations in a regime mostly began at an early stage. In more than 75 per cent of the regime elements IGOs were identified as organizations which were actively involved in political processes. IGOs provided expertise required for monitoring and scientific research or for the development of policies in regimes. They frequently stimulated implementation of international research programs and caused international and national research networks to collaborate in epistemic processes. For example, the evolution of knowledge in the Danube river protection regime has strongly benefited from research and policy initiatives of intergovernmental organizations like the World Bank, UNDP, the Global Environmental Facility or the European Union (Kaspar 1999, 188f). A similar role of intergovernmental organizations has been identified for regimes managing, inter alia, imports and exports of hazardous waste, the protection of biodiversity, global climate, or stratospheric ozone. One of the most prominent examples is the Intergovernmental Panel on Climate Change (IPCC) which has been established in the late 1980s under the auspices of WMO and UNEP and whose working groups contributed to improve knowledge about the causes and impacts of, or about policies and adaptation strategies required for responding to, climate change. The evolution of dense networks between IGOs, semi-governmental or private research organizations could frequently be observed in environmental regimes. IIASA's RAINS model which informed about the options available for developing sulfur protocols under the umbrella of the LRTAP-Convention has been an example for a successful partnership in a regime with a semi-governmental organization. UNEP and other semi-governmental or private non-state actors closely collaborated in monitoring the depletion of stratospheric ozone or in assessing technical and economic options for the phase-out of ozone-depleting substances (Clark, van Eijndhoven, Dickson et. al. 2001, 50–53).

The nation-state influences and determines decision-making in inter-governmental organizations about the character of research programs. Nation-states are aware that independent action by a single state will not lead to desired results or will be more costly and less effective than collaborative research efforts. IGOs are dependent on governmental priorities and are closely linked with the work of state-financed national research facilities. Their efforts can partly be understood as a result of governmental strategies which intend to improve the legitimacy of governmental policies. However, research programs carried out by international organizations can develop their own momentum. When scientists produce proof of cause-effect relationships in environmental issue areas or when policy experts reach consensus about the feasibility of policy options, governments have to consider these findings and come under pressure to adapt their own policies accordingly.

National or international scientific organizations accompanied and strengthened efforts by IGOs. It would be misleading to conceive of the evolution of knowledge in international regimes purely as a result of the work and programs led by international organizations. The finding about the role played by scientific and technical non-state actors during the political process is supported by other data. They indicate for 149 of the 172 regime elements that science-based communities which represent networks that are located in national or international research institutes, in intergovernmental organizations or national research departments were present and active during political processes. However, epistemic communities as a very coherent type of network have obviously played a less influential role during political processes than it is maintained by adherents of the concept on epistemic communities. Empirical data suggest the conclusion that the creation of new knowledge is normally not the result of a coherent transnational group of scientific and technical experts, but emerges from the activities of a broad range of actors.

The achievement of improved levels of understanding about the nature of the problem or of information about policy options coincides with long-term activities of science-based non-state actors. This suggests the conclusion that these actors could contribute to reduce uncertainties related to the cognitive setting. Data also reveal that many scientific and technical non-state actors were engaged in public discourse of regimes where they made efforts to interpret the relevance of scientific knowledge or of policy proposals. Such efforts were not only made by IGOs like the World Bank, UNEP, UNDP or FAO which have been identified in a number of regimes as organizations which influenced political processes with ideational or entrepreneurial forms of influence. Data also illustrate that non-state actors other than intergovernmental organizations were important for the production and dissemination of scientific knowledge or for developing and politicizing policy options. For the most part, the activities of these non-state actors were limited to single issue-areas. Scientific or activist networks like the Aktionskonferenz Nordsee (AKN), the Antarctic and Southern Ocean Coalition (ASOC), the Basel Action Network, the Coalition Clean Baltic, the Danube Forum, the International Association for Great Lakes Research (IAGLR), or the Scientific Committee on Antarctic Research (SCAR), to mention only a few, were identified as actors which influenced the development of new knowledge and policy options or which contributed to disseminating this knowledge to the transnational public. In addition, similar forms of ideational and entrepreneurial influence have been identified for big non-state actors like Greenpeace, WWF, or Friends of the Earth. The activities of these transnational nonprofit activist interest groups were observed in more than one regime. For a number of prominent non-state actors which are usually subsumed under the category of advocacy actors, ideational and entrepreneurial forms of influence were identified in many regimes.

Most of the important individuals identified in single regimes used ideational or entrepreneurial forms of influence. These individuals belonged to governmental, semi-governmental or fully private organizations and contributed as policy entrepreneurs or scientists to the evolution of the cognitive setting. In particular, individual members of international organizations, regime secretariats, or non-profit environmental activist interest groups have been referred to by coding case study experts as actors which initiated or supported the establishment of new research programs or the development of new policies. For example, legal experts like Cyrille de Klemm influenced the work of nature conservation regimes and launched initiatives for the development of international conservation policies. The work of marine scientist Jacques Cousteau who produced several dozens of documentary films about the diversity and depletion of the marine environment influenced the perception of marine issues in the broader public and by governments. Ideational influence has been ascribed to Jacques Cousteau in particular for the evolution of the MARPOL regime and the London Convention regime.

Our data show that participation by industrial organizations or by affected economic industries in regime politics increased over time. But this does not mean that policy dialogue between states, non-profit organizations and business actors has significantly intensified in environmental regimes. For a broad number of regimes, economic actors have been coded in regard to the forms of influence displayed by these organizations during political processes. This group consists of actors like the Alliance for Responsible CFC Policy, Dupont, Industrial Chemical Industries (ICI), the Council of Great Lakes Industries, fishing associations and fishing industries from various countries, the International Association of Antarctica Tour Operators (IAATO) or the International Chamber of Shipping. Obviously, inclusion of these actors in the process of policy formulation can contribute to assess the feasibility or to improve the quality of policies agreed upon in regimes. Progress in the consensual knowledge about cause-effect-relationships or about technical and economic solutions can involve that economic actors are willing to change their attitudes towards environmentally-sound production technologies. Industrial Chemical Industries (ICI) and Dupont are two – though rare – cases where advanced knowledge has finally caused a change in the behavior of these chemical firms from opponents to supporters of a phase-out of ozone-depleting substances. Growing impact of environmental regimes on market conditions caused a general increase in participation by economic actors. But some examples included in the database reveal that the debate about policy options has often been determined by more or less strong conflict between environmental and economic interests. For example, a significant amount of economic actors whose activities were coded for the database played the role of laggards which aimed to water down regime policies. Other firms or industrial associations took on more environment-friendly attitudes and supported the development of environmental policies. Inclusion of economically affected actors was often prerequisite for developing feasible policies.

The above-described findings suggest the conclusion that the reduction of uncertainty as one among a number of grounds for legitimacy could frequently be achieved because non-state actors contributed scientific expertise and resources or participated in developing policies which could be used for managing environmental problems. In many cases, the state responded to new scientific knowledge produced by non-state actors and intensified the establishment of national and international research programs. The state itself often launched initiatives for the creation of new knowledge or for the development of policies because it has traditionally been responsible for the establishment and financing of research infrastructure on domestic and international levels. These findings illustrate that the nation-state remained deeply involved in the evolution of the cognitive setting.

Conclusion

Without the existence of various programmatic activities it would have been far more difficult – and often impossible – to improve the understanding about the nature of the problem or to raise the completeness of information about policy options. Measurement of our dependent variable led to the result that these uncertainties have been reduced more or less significantly during the existence of most of the regimes explored. These findings support theoretical assumptions held by those students of international governance who allude to the role of regimes as producers of consensual knowledge and as facilitators of learning (Ernst B. Haas 1990; Peter M. Haas 1992). Networks of scientific and technical experts have been identified as being actively involved in political processes for most of the regimes explored. The creation of knowledge often emerges from complex interaction of various actors. Empirical findings support the assumption of functional theories according to which regimes provide an opportunity for reducing transactions costs and for improving communication and the exchange of information (Keohane 1984). The observed changes by states in their understandings about the nature of the problems or concerning the completeness of information about policy options can be considered as one among a number of factors which contribute to achieve changes in state preferences. This finding suggests that an important factor for the legitimacy of international regimes arises from the way how the institution will be designed and whether the institutional design will guarantee that a regime can contribute to reduce various uncertainties related to understandings about the nature of the problem or about the completeness of information about policy options (Breitmeier 2006). Obviously, the institutional design of environmental regimes increasingly caused the proliferation of this ground for legitimacy, even though the complexity of many problems involves that improvements can often be accomplished only after a lengthy process.

It became obvious that changes in the cognitive setting could not be achieved without the support by non-state actors, regardless whether these actors were intergovernmental organizations, state-financed research organizations, non-profit-oriented actors, or economic firms. Non-state actors contributed to the reduction of uncertainties and to promoting one among a number of grounds which justify obedience in world politics. Not each non-state actor is able to contribute to improvements in the cognitive setting equally. Such contribution can primarily be made by actors with scientific and technical expertise. The creation and further evolution of knowledge is very much determined by collaboration rather than antagonism between international society and emerging global civil society. Political conflict between these two worlds is more likely to occur during the development of policy options, whereas monitoring or scientific research normally take place in a more de-politicized environment. Non-state actors can not make these contributions without the existence of various programs which allow them to provide their expertise or to participate in the formulation of policies. It seems plausible to conclude that state-dominated regimes have increasingly benefited from the potential of non-state actors in the creation of knowledge or in developing policies for the management of trans-boundary environmental problems.

Chapter 7
Regimes and Compliance

Compliance is a ground that justifies obedience by an actor particularly when others comply. Non-compliance by important members creates a situation where complying states contribute to the production of a collective good but will be exploited by free-riders (Olson 1965, 29). Non-compliance by states which are important for international problem-solving can diminish the belief on the part of complying nations that they will be able to achieve their goals in an institution. It can undermine the regularity of social relations and cause an erosion of social order. Whether norms and rules deserve obedience depends not only on compliance by others. The character of norms and rules also determines whether compliance is justified. Environmental norms and rules deserve obedience if they give prospect of improvements in the environment and if they avoid that strongly asymmetrical outcomes will emerge. In the following, attention will *first* be paid to measuring the level of compliance by all members in the issue area. Measurements will also focus on the compliance behavior of important states. It will be clarified whether deepened penetration of states by regime policies increased compliance problems. In exploring the causal impact of regimes for observed levels of compliance, it will be assessed whether compliance resulted primarily from the operation of international regimes or has been caused by external factors. In presenting the empirical results collected for our dependent variable, the frequency distribution on ordinal scales which were developed for measuring the degree of compliance by members will be illustrated for all regime elements and for a sub-set of regime elements ending in 1998. In addition, a check for intercoder reliability will be carried out for data of our dependent variable. In addition, those events and actions will be explored that were necessary or significant for translating international commitments into domestic obligations in important member states. In a *second* step, it will be explored whether institutional mechanisms or non-state actors contributed to compliance. A theoretical approach focusing on the institutional design as a factor that impacts on compliance leads us to expect that compliance is dependent on the existence of various institutional mechanisms. Since environmental governance systems frequently lack mechanisms which could be used to enforce regime rules in case of non-compliance, special attention will be given to exploring whether institutional mechanisms emerged which could influence the management of compliance problems. An approach which focuses on the role of non-state actors suggests that compliance problems will partly be diminished by the involvement of governments and stakeholders in transnational discourse about the necessity and rightness of policies. This approach leads us to expect that governments will be forced by environmental actors to explain the causes for noncompliance to the public or be pressurized to change behavior accordingly. In addition, this approach leads us to expect that compliance by states is dependent on the service and expertise of non-state actors which they contribute in compliance mechanisms and on the willingness of affected economic groups to implement regime policies.

Compliance

Did members comply with regime rules? Did regimes have a causal impact for achieving observed levels of compliance? Regime elements will be the major unit of analysis in exploring compliance behavior by states. The focus of analysis on regime elements makes it possible to measure conformity with norms and rules during time periods. The dichotomy between compliance and non-compliance provides a starting-point for developing more fine-grained measurements which can be used to illustrate various dimensions of compliance behavior by states. The pure distinction between compliance and non-compliance ignores the fact of over-compliance which occasionally determines the behavior of single states. The scales used for measuring state behavior reflect that various forms of compliance or noncompliance can be found. An ordinal scale from one to five is used for measuring compliance. Behavior which exceeds regime requirements and behavior which does not conform at all have been determined as upper and lower limits on this scale. The upper limit of this ordinal scale considers the fact of over-compliance by states whose behavior demonstrates that implementation of more far-reaching measures is possible. The lower limit of this ordinal scale refers to behavior where actors do not at all conform with regime rules to any significant or important degree. The actual compliance behavior required to meet obligations of international regimes is established one value below the level of over-compliance. Two other forms of non-compliance on our ordinal scale refer to behavior where actors comply with some requirements but not all or where behavior conforms some (but not all) of the time and/or to some degree but not completely.

Compliance by all Members

Data about the compliance behavior at the level of all regime members are available for 130 regime elements. They cover 22 of our total of 23 regimes. For 80 of these 130 regime elements behavior meets regime requirements (62) or is determined by over-compliance (18). Thus, more than 60 per cent of the data collected for these 130 regime elements reflect compliance behavior that lives up to the rules established by regimes (see Table 7.1). No other finding arises if only those 74 regime elements will be analyzed which end in 1998. By and large, full compliance or over-compliance can be found in more than half of the regime elements. The majority of data are distributed on the scale which indicates that behavior meets regime requirements. For some regime elements, over-compliance

Table 7.1   **Compliance with Regime Provisions (Level of All Regime Members)**

Regime Elements		41 pairs of regime elements (codings at most differing by one value)*		23 pairs of regime elements which end in 1998 (codings at most differing by one value)*		Compliance by all Member States
18	(13.8%)	6	(7.3%)	4	(8.7%)	Behavior exceeds regime requirements
62	(47.7 %)	45	(54.9%)	26	(56.6%)	Behavior meets regime requirements
34	(26.2%)	28	(34.1%)	14	(30.4%)	Behavior conforms with some requirements but not all
13	(10.0%)	3	(3.7%)	2	(4.3%)	Behavior conforms some of the time and/or to some degree but not completely
3	(2.3%)	0	(0.0%)	0	(0.0%)	Behavior does not conform at all
130	100.0 %	82	100.0%	46	100.0%	Total

* Two codes exist for each regime element. Pairs of regime elements were omitted, if disagreement between coding experts has reached more than one value on the scale. Accordingly, the subset of 41 pairs of regime elements ignores 9 pairs (or 18 data). The subset of 23 pairs applying to regime elements which end in 1998 ignores 4 pairs (or 8 data). Percentage figures were rounded up and down.

has been identified. The total of data also illustrates that environmental regimes are frequently determined by compliance problems. These problems can be more or less severe in single cases. The vast majority of data reflecting behavior that does not meet regime requirements refers to behavior where states conform with some requirements but not all or where behavior conforms some of the time or to some degree but not completely. Complete non-compliance occurred very rarely. Even though these data indicate that more or less insufficient levels of compliance could be found in many regime elements, they also show that progress was made if efforts to manage these compliance problems effectively had been strengthened. In some regimes where watersheds distinguished two or more time periods, levels of compliance improved during more recent periods.

A sub-set of 100 data can be checked for intercoder reliability. A rather satisfactory level of intercoder reliability can be found. Identical answers were given for 25 pairs of regime elements, whereas for 16 pairs codings differed by one value on our scale from one to five. Data provided by two experts varied by more than one value for 9 pairs of regime elements. Forty-one pairs of regime elements represent a particularly reliable data set. Accordingly, for more than 80 per cent of the regime elements coded by two experts data about compliance of all members differed by no more than one value on the measurement scale. The 41 pairs of regime elements refer to 14 regimes. More than 50 per cent of the data included in this subset of 41 pairs again indicate compliance or over-compliance by all members. A similar finding emerges for the sub-set of pairs of regime elements which end in 1998.

Which findings can be gained from this sub-set of 41 pairs of regime elements for the level of single regimes? For the Antarctic regime or for the ozone layer regime, sufficient levels of compliance have been identified for nearly all regime elements. For the Barents Sea fisheries regimes which includes only a single regime element, agreement existed between coding experts that state behavior meets requirements. The river regimes for the protection of the Rhine or the Danube and the Great Lakes management regime were mainly determined by compliance by members, but in some regime elements at least one of the data indicate that behavior meets only with some but not all regime requirements. Our data on compliance refer mainly to legally binding international arrangements. But the 1987 Rhine Action Plan which has been developed by members of the Rhine regime to improve the ecosystem of the Rhine can be used for illustrating that states can also comply with non-binding instruments.[1] For the whaling regime, improvements by one value to behavior that meets regime requirements occurred during the period from 1982 to 1998, whereas an a bit lower level of compliance existed between 1946 and 1982. For the Baltic Sea regime coding experts agreed that one regime element was determined by compliance problems between 1974 and 1992. This regime is one of the few cases where more or less strong disagreement existed between both coding experts about the degree of compliance that could be found in single regime elements. Some elements of the climate regime are determined by compliance. But a minor difference between coding experts on whether behavior meets requirements or whether it meets only some but not all rules also exist for some elements of this regime. This type of minor differences could also be found for the data provided for the CITES-regime. One expert indicated for this regime that behavior met requirements whereas the other expert made an a bit less optimistic assessment.

Similar minor differences existed for those regime elements of the North Sea regime which end in 1998. For other elements of this regime experts agreed that compliance behavior by all states reflects conformity with some but not all rules. These data indicate that some regimes in fact experience compliance problems on a more or less regular basis. But these compliance problems must not inevitably

---

1   It is often difficult to distinguish between hard law and soft law. For example, "recommended measures" adopted by the Antarctic Treaty Consultative Meeting (ATCM) expanded Treaty obligations over time. More than 200 recommended measures and more than fifty associated instruments were adopted by ATCMs by 1999 although their status as legally binding norms is unclear and they "may be viewed as non-binding norms, or they may be seen to constitute hard law, or they may be considered as something in between" (Joyner 2000, 164).

be so severe that they threaten the existence of a regime. The minor differences that exist among the codings of some case study analysts illustrate that it is sometimes difficult to draw the exact borderline between full compliance and less significant cases of noncompliance. These difficulties are also inevitable if compliance behavior is assessed for a long-term period. The severity of many compliance problems can not be neglected. But they frequently represent a normal situation which members intend to change in the mid- or long-term. For example, experts agreed for the Ramsar regime on wetlands that behavior did not fully meet regime requirements. But the regime is determined by constant efforts to improve the implementation of the Ramsar Convention on wetlands at the global level. The London Convention regime can be referred to as another example for occasional failure to achieve a complete level of compliance at the global level. Noncompliance is limited to a smaller number of members. Coding experts also agreed that regulations of this regime that referred to wastes and substances the dumping of which is either prohibited or which, in principle, may be dumped were not fully complied with by all members and that conformity of behavior occurred with some but not all rules. These findings apply to both periods from 1972 to 1991 and from 1991 to 1998 equally. A rather low level of compliance has been found in the desertification regime. Agreement existed between both experts that behavior conforms some (but not all) of the time and/or to some degree but not completely. At least some of the compliance problems occurring in this regime may have been caused by the fact that the United Nations Convention to Combat Desertification was a new regime. Not enough time and institutional energy may have been available to manage compliance problems effectively.[2]

For three regimes data about compliance by all members are available from only one expert. For the hazardous waste regime, behavior that meets regime requirements has been identified for four regime elements coded. The regime managing intentional oil pollution at sea has failed to achieve full compliance by all members. But conformity with the MARPOL-provisions which existed since 1973/78 has improved compared to compliance with regulations of OILPOL in the period between 1954 and 1973/78 (Mitchell 1994, 204–10). As far as MARPOL and various regional memorandums of understanding are concerned, behavior of all members complied with some but not all provisions. Regulations that emerged for the general management of fisheries in the South Pacific region have been complied with in former and more recent periods of this regime. The element established for managing compliance issues in this regime has been determined by compliance of all members after an interim period needed between 1979 and 1982 to set up institutional provisions which could guarantee proper functioning of this element.

Did growing density and specificity of regime rules impact on compliance behavior? Even though a significant number of regime elements were still far from

2   The Convention which was adopted in June 1994 entered into force in December 1996 only two years before our examination period ended.

achieving sufficient levels of compliance at the end of the twentieth century, the encouraging news is that neither the expansion of the functional scope of regime rules nor increasing density and specificity of rules affected compliance by states negatively. The IRD includes supplementary data about the functional scope or the density and specificity of regime rules. These dimensions were measured on two separate ordinal scales that reach from one to five. One of these two ordinal scales considered a very broad or very narrow functional scope as upper and lower limits. The functional scope was conceived of as very broad if the regime covered all important issues considered necessary for inclusion in the regime. A regime had a very narrow functional scope if its rules covered only a limited number of issues that were considered necessary for managing problems effectively. Similar upper and lower limits delimit the other ordinal scale used for measuring the degree of density and specificity of regime rules. If a regime comprised a very comprehensive set of rules and/or established rules that are rather strong, it has been conceived of as very deep. If there was only a very limited number of rules compared to the density of rules which was necessary for effective management and/or rules were rather weak compared to the specificity of rules considered necessary for effective problem-solving, the regime was defined as very shallow.

The findings that are included in Table 7.2 indicate that a broad or very broad functional scope of regime rules existed in 39 of the 74 regime elements where compliance or over-compliance has been measured. In 33 of the 49 regime elements where compliance levels were not sufficient, the functional scope has also been broad or very broad. Obviously, a broad functional scope does not inevitably lead to weaker level of compliance. On the other hand, there is a significant number of regime elements where such weak levels of compliance correlate with a broad or very broad functional scope of regime rules. On the first view, findings in Table 7.3 give the impression that compliance behavior is more likely to occur in cases where the density and specificity or regime rules is underdeveloped. The density and specificity of rules has been at most at a medium level in 50 of the 75 regime elements where compliance levels have been satisfactory. But a detailed view on the remaining 25 regime elements of this group reveals that the broad majority of these cases (23) ended only in 1998. The type of cases that combines satisfactory compliance levels and the existence of relatively deep and dense rules emerged primarily in the late 1980s and in the 1990s. It is also worth to look on a subset of those 50 cases where levels of compliance are quite high but where the density and specificity of rules is relatively shallow or at most at medium levels. Twenty-six of these 50 regime elements exist predominantly before the 1990s and end in 1994 at the latest. This finding supports the conclusion that the broadening of the functional scope and the deepening of the density and specificity of regime rules did not negatively impact on compliance behavior.

A case in point is the ozone regime. A medium or broad functional scope of regime rules and deep density and specificity of rules coincided with full compliance by regime members. In addition, single regime elements were determined by over-compliance. A broadening of the functional scope and deepening in the density and

Table 7.2   Correlation between Functional Scope of Regimes and Compliance (Level of Regime Members)

| Total of Regime Elements | | Very narrow | | Functional Scope of Regime Rules | | | | | | | | Compliance by all Member States |
				Narrow		Medium		Broad		Very broad		
17	(13.8%)	4	(30.8%)	0	(0.0 %)	4	(15.4%)	8	(17.4%)	1	(3.8%)	Behavior exceeds regime requirements
57	(46.3%)	6	(46.1%)	9	(75.0%)	12	(46.2%)	16	(34.8%)	14	(53.9%)	Behavior meets regime requirements
34	(27.7%)	0	(0.0%)	3	(25.0%)	8	(30.8%)	13	(28.2%)	10	(38.5%)	Behavior conforms with some requirements but not all
12	(9.8%)	2	(14.4%)	0	(0.0%)	1	(3.8%)	8	(17.4%)	1	(3.8%)	Behavior conforms some of the time and/ or to some degree but not completely
3	(2.4%)	1	(7.7%)	0	(0.0%)	1	(3.8%)	1	(2.2%)	0	(0.0 %)	Behavior does not conform at all
123	100.0%	13	100.0%	12	100.0%	26	100.0%	46	100.0%	26	100.0%	Total

Table 7.3   Correlation between Density/Specificity of Rules and Compliance (Level of All Regime Members)

| Total of Regime Elements | | Very shallow | | Density/Specificity of Regime Rules | | | | | | | | Compliance by all Member States |
				Shallow		Medium		Deep		Very deep		
18	(14.4%)	1	(5.9%)	2	(8.7%)	6	(14.3%)	8	(23.5%)	1	(11.1%)	Behavior exceeds regime requirements
57	(45.6%)	8	(47.1%)	7	(30.4%)	26	(61.9%)	12	(35.3%)	4	(44.4%)	Behavior meets regime requirements
34	(27.2%)	4	(23.5%)	7	(30.4%)	8	(19.0%)	12	(35.3%)	3	(33.3%)	Behavior conforms with some requirements but not all
13	(10.4%)	3	(17.6%)	7	(30.4%)	1	(2.4%)	1	(2.9%)	1	(11.1%)	Behavior conforms some of the time and/ or to some degree but not completely
3	(2.4%)	1	(5.9%)	0	(0.0%)	1	(2.4%)	1	(2.9%)	0	(0.0%)	Behavior does not conform at all
125	100.0%	17	100.0%	23	99.9%	42	100.0%	34	99.9%	9	99.9%	Total

specificity of regime rules occurred also in the component of the Antarctic regime on the conservation of Antarctic marine living resources. The broadening and deepening of regime rules came along with the achievement of full compliance by all members during the more recent period in the 1990s. The preceding period of this component in the 1980s was determined by a little lower level of compliance.

The question remains on whether these findings already disprove the thesis according to which compliance problems will increase with the depth of cooperation needed for problem-solving (Downs, Rocke and Barsoom 1996). One could argue in favor of this hypothesis that most of the cases explored have not yet reached the kind of depth that leads us to expect the emergence of severe compliance problems. In some regimes which focus on the management of truly comprehensive problems like climate change or desertification implementation of more comprehensive international policies will be required for the attainment of environmental goals. The acid test in regard to whether the depth of cooperation impacts on the willingness of capacities of states to comply is still imminent for many environmental regimes. The broadening and deepening of rules did not inevitably lead to lower levels of compliance. This finding suggests that supportive factors like the institutional provisions for compliance management or political backing for compliance can contribute to reducing the constraints for compliance arising from the depth of cooperation. On the other hand, the above-described findings illustrate that compliance problems can occur also in regimes where the functional scope or the density and specificity of regime rules hardly correspond to the depth of cooperation that is supposed to make noncompliance very likely. The Ramsar regime on wetlands experienced occasional compliance problems although it is still a shallow regime.

Which causal impact did regimes have on compliance at the level of all members? The causal influence of a regime on the conformance of members has been measured on a scale from one to four. When a regime had a little or no causal impact, non-regime factors were considered to mainly account for the state of the world. Economic recession, technological innovation, or changing consumer patterns are external factors which can contribute to achieving environment-friendly behavior by states without that states will have to implement measures to the degree normally required for complying with environmental norms.[3] A modest causal impact involved that the regime matters with regard to observed levels of compliance, but non-regime factors were considered more important. A large causal influence was assumed if a regime accounted equally with non-regime factors for outcomes or has proven to be more important than non-regime factors.

---

3   For example, Jorgen Wettestad (1999) observes a rather high level of compliance by states with the regulations of the LRTAP-regime that provides for emission reductions of single air pollutants in Europe. However, the author also concludes that for the UK, the Netherlands, and Norway "the reductions and compliance levels (...) are apparently explained by processes that are not primarily related to environmental protection", but were caused by industrial recession and changes in the energy-mix (Wettestad 1999, 91).

A fourth scale has been used to determine a negative influence of a regime toward conformance with requirements. Such a negative impact of regimes on compliance by all members has been identified in none of the regime elements. Hundred and twenty data illustrate the causal impact of an institution for compliance in regime elements (see Table 7.4). These data indicate that behavior of all members at least met with regime rules in more than 60 per cent of the regime elements (80 of a total of 130). For the majority of regime elements a large causal influence has been ascribed to a regime for achieved levels of compliance. Regimes were ascribed a large causal influence in 52 of the 80 cases in which compliance (62) or over-compliance (18) prevailed. The regime impact was large in nearly each of the cases where behavior of all members exceeded requirements of regime rules.

Compliance by Important States

Did important states comply with norms and rules? The majority of data collected about compliance behavior of more than 50 important states indicates that behavior meets or exceeds regime requirements. Between four and seven important states were coded in a regime element regarding their compliance behavior. The findings which correlate compliance behavior of important states with the causal impact of regimes are included in Table 7.5. The United States has been coded nearly a 100 times because of its relevance as a member in a large number of regimes. Germany, the United Kingdom, or the Russian Federation and the former Soviet Union were also frequently coded. In 92 instances correlation can be made between the compliance behavior of the United States and the causal impact of regimes. Of the 79 data indicating that behavior of the United States met (54) or exceeded (25) regime requirements, regimes had a large causal impact for this behavior in 57 cases. These data illustrate that the behavior of the US was mainly determined by compliance and that regimes were important factors for this behavior in more than 70 per cent of the cases. A drop of bitterness may exist if one considers weak compliance by the US with norms and rules of the desertification regime by the end of the twentieth century. But such a clearly negative assessment has been made by coders with respect to the behavior of this country for no other regime.[4] The large causal impact ascribed to environmental regimes for compliance behavior by the United States suggests the conclusion that even a powerful country prefers the existence of institutional mechanisms which contribute to reducing uncertainty about compliance behavior by others. The significant causal role ascribed to regimes in this case suggests that institutional mechanisms played an important role for stimulating implementation of various measures on the domestic level.

---

4   The United States ratified this convention only in late 2000 and it entered into force for this country in early 2001. Thus, weak compliance during the 1990s has been caused by the fact that this country has not formally acknowledged the obligations of this convention during this period.

Table 7.4 Causal Impact of Regimes on Compliance (Level of All Regime Members)

Compliance by all States	Total of Regime Elements		Causal Impact of Regimes				
		Not applicable	Little or no causal impact	Modest causal influence	Large causal influence	Negative causal influence
Behavior exceeds regime requirements	18 (13.8%)	1 (10.0%)	0 (0.0%)	0 (0.0%)	17 (21.3%)	0 (0.0%)
Behavior meets regime requirements	62 (47.7%)	4 (40.0%)	3 (37.5%)	20 (62.5%)	35 (43.8%)	0 (0.0%)
Behavior conforms with some requirements but not all	34 (26.2%)	2 (20.0%)	3 (37.5%)	8 (25.0%)	21 (26.3%)	0 (0.0%)
Behavior conforms some of the time and/or to some degree but not completely	13 (10.0%)	3 (30.0%)	1 (12.5%)	2 (6.3%)	7 (8.8%)	0 (0.0%)
Behavior does not conform at all	3 (2.3%)	0 (0.0%)	1 (12.5%)	2 (6.3%)	0 (0.0%)	0 (0.0%)
Total	130 (100.0%)	10 100.0%	8 100.0%	32 100.1%	80 100.2%	0 0.0%

Table 7.5 Causal Impact of Regimes on Compliance of Important States

Total of Regime Elements	Little or no causal impact	Moderate causal impact	Large causal impact	Negative causal impact	Level of Compliance*	Important Nation
11	0 (0.0%)	0 (0.0%)	11 (100.0%)	0 (0.0%)	1	Australia
18	0 (0.0%)	9 (50.0%)	9 (50.0%)	0 (0.0%)	2	
1	0 (0.0%)	1 (100.0%)	0 (0.0%)	0 (0.0%)	3	
0	0 (0.0%)	0 (0.0%)	0 (0.0%)	0 (0.0%)	4	
0	0 (0.0%)	0 (0.0%)	0 (0.0%)	0 (0.0%)	5	
7	0 (0.0%)	0 (0.0%)	6 (85.7%)	1 (14.3%)	1	Canada
17	0 (0.0%)	6 (35.3%)	11 (64.7%)	0 (0.0%)	2	
3	0 (0.0%)	0 (0.0%)	3 (100.0%)	0 (0.0%)	3	
0	0 (0.0%)	0 (0.0%)	0 (0.0%)	0 (0.0%)	4	
1	1 (100.0%)	0 (0.0%)	0 (0.0%)	0 (0.0%)	5	
2	0 (0.0%)	0 (0.0%)	2 (100.0%)	0 (0.0%)	1	China
11	2 (18.2%)	2 (18.2%)	7 (63.6%)	0 (0.0%)	2	
2	0 (0.0%)	2 (100.0%)	0 (0.0%)	0 (0.0%)	3	
0	0 (0.0%)	0 (0.0%)	0 (0.0%)	0 (0.0%)	4	
0	0 (0.0%)	0 (0.0%)	0 (0.0%)	0 (0.0%)	5	
8	0 (0.0%)	0 (0.0%)	8 (100.0%)	0 (0.0%)	1	European Union
12	1 (8.3%)	5 (41.7%)	6 (50.0%)	0 (0.0%)	2	
3	0 (0.0%)	2 (66.7%)	1 (33.3%)	0 (0.0%)	3	
1	0 (0.0%)	0 (0.0%)	1 (100.0%)	0 (0.0%)	4	
0	0 (0.0%)	0 (0.0%)	0 (0.0%)	0 (0.0%)	5	
22	0 (0.0%)	10 (45.5%)	12 (54.5%)	0 (0.0%)	1	Germany
20	4 (20.0%)	10 (50.0%)	6 (30.0%)	0 (0.0%)	2	
9	0 (0.0%)	9 (100.0%)	0 (0.0%)	0 (0.0%)	3	
0	0 (0.0%)	0 (0.0%)	0 (0.0%)	0 (0.0%)	4	
0	0 (0.0%)	0 (0.0%)	0 (0.0%)	0 (0.0%)	5	

Table 7.5 Continued

Total of Regime Elements	Causal Impact of Regimes				Level of Compliance*	Important Nation
	Little or no causal impact	Moderate causal impact	Large causal impact	Negative causal impact		
2	0 (0.0%)	0 (0.0%)	2 (100.0%)	0 (0.0%)	1	India
9	2 (22.2%)	2 (22.2%)	5 (55.6%)	0 (0.0%)	2	
4	0 (0.0%)	4 (100.0%)	0 (0.0%)	0 (0.0%)	3	
0	0 (0.0%)	0 (0.0%)	0 (0.0%)	0 (0.0%)	4	
0	0 (0.0%)	0 (0.0%)	0 (0.0%)	0 (0.0%)	5	
3	0 (0.0%)	0 (0.0%)	3 (100.0%)	0 (0.0%)	1	Japan
16	0 (0.0%)	5 (31.3%)	11 (68.8%)	0 (0.0%)	2	
6	0 (0.0%)	1 (16.7%)	5 (83.3%)	0 (0.0%)	3	
1	0 (0.0%)	1 (100.0%)	0 (0.0%)	0 (0.0%)	4	
2	1 (50.0%)	0 (0.0%)	0 (0.0%)	1 (50.0%)	5	
16	0 (0.0%)	0 (0.0%)	16 (100.0%)	0 (0.0%)	1	United Kingdom
22	3 (13.6%)	11 (50.0%)	8 (36.4%)	0 (0.0%)	2	
3	0 (0.0%)	0 (0.0%)	3 (100.0%)	0 (0.0%)	3	
0	0 (0.0%)	0 (0.0%)	0 (0.0%)	0 (0.0%)	4	
0	0 (0.0%)	0 (0.0%)	0 (0.0%)	0 (0.0%)	5	
25	1 (4.0%)	0 (0.0%)	22 (88.0%)	2 (8.0%)	1	United States
54	4 (7.4%)	15 (27.8%)	35 (64.8%)	0 (0.0%)	2	
7	1 (14.3%)	2 (28.6%)	4 (57.1%)	0 (0.0%)	3	
3	0 (0.0%)	3 (100.0%)	0 (0.0%)	0 (0.0%)	4	
3	2 (66.7%)	0 (0.0%)	0 (0.0%)	1 (33.3%)	5	
13	1 (7.7%)	0 (0.0%)	12 (92.3%)	0 (0.0%)	1	USSR/Russian
27	2 (7.4%)	5 (18.5%)	20 (74.1%)	0 (0.0%)	2	Federation
3	1 (33.3%)	0 (0.0%)	2 (66.7%)	0 (0.0%)	3	
13	1 (7.7%)	7 (53.8%)	3 (23.1%)	2 (15.4%)	4	
5	5 (100.0%)	0 (0.0%)	0 (0.0%)	0 (0.0%)	5	

*Level of Compliance: 1=Behavior exceeds regime requirements; 2=Behavior meets regime requirements; 3=Behavior conforms with some requirements but not all; 4=Behavior conforms some of the time and/or to some degree but not completely; 5=Behavior does not conform at all.

A more ambiguous finding arises concerning compliance by Russia or the former Soviet Union. From the total of seven regimes for which compliance behavior of Russia or the former Soviet Union has been determined, in six regimes relatively satisfying or full levels of compliance by Russia can be found. But these levels of compliance existed normally not for each regime element. An exception is the Barents Sea fisheries regime where full compliance by Russia prevailed. In some regimes, compliance behavior improved over time. More or less significant compliance problems by Russia were identified for single elements of the ozone regime during the 1990s. Compliance problems also occurred in the whaling regime during its early period between 1946 and 1982. But behavior improved since then to full compliance. A similar dynamic characterized Russian compliance with CCAMLR-regulations in the Antarctic regime. During the 1980s, behavior complied with some but not all rules. But it improved to full compliance for the period between 1989/91 and 1998. All other elements of the Antarctic regime were determined by full compliance by Russia since the regime emerged in 1959. For the Antarctic regime, the Barents Sea fisheries regime, the whaling regime or various elements of the London Convention regime mainly a large causal impact of the regime has been identified for the degree of Russian compliance with regime rules.

It is possible to make correlation between Germany's compliance behavior and the causal impact of regimes for 51 data. Forty-two data indicate that behavior exceeded (22) or met (20) regime requirements. Regimes had mostly a moderate (20) or large causal impact (18) on Germany's compliance behavior. On balance, the relatively favorable picture that emerges in regard to compliance behavior of industrialized countries can be confirmed with data collected for countries like the United Kingdom, Australia, Canada, or Japan. Compliance behavior for each of these countries has been measured for at least two regimes. For a total of 119 data correlations can be made between compliance behavior of these countries and the impact of regimes. Ninety data indicate that behavior of these countries reflected compliance. In 75 of the 90 cases, regimes had a large causal impact on compliance. The small rest of data primarily concerns less serious forms of non-compliance. Similar findings arise from data collected for six regimes about the compliance behavior of the European Union. A good compliance record arises for countries like China or India. Twenty-three of the 30 data indicate that behavior of these countries met or exceeded regime requirements. Regimes had a large causal impact on compliance in 16 cases.

Politics and Implementation on the Domestic Level

Political battles on the domestic level over the terms of environmental agreements can be so severe that they delay or prevent ratification or implementation of other steps which are necessary for acceding to an international agreement. Broad variance can exist among member countries in regard to whether they experienced such political battles. The severity of political battles that occurred on the domestic level of an important state could be coded on a scale from one to five. Very strong political battles or not at all existing political battles were used as upper and lower limits on that scale. In combination with the coding of the severity of political battles, another variable which included scales from one to four was used for exploring how many states were affected by these conflicts. Some countries (less than 20 per cent of regime members) and all countries (all regime members) were used as upper and lower limits on that scale.

During the process of data collection it became obvious that our knowledge about the political processes determining the translation of international environmental norms into domestic obligations is often underdeveloped as far as single regimes are concerned. Two observations that can be made on the basis of our data support this conclusion: First, a proportionally large number of experts avoided to submit information about the degree to which political battles over the terms of an agreement could be observed on the domestic levels of regime members; second, major disagreements could be observed more frequently between those data which could be checked for intercoder reliability. Nevertheless, for a number of regimes findings based on two independent codings reveal that the whole range of political battles included on our ordinal scale could be found during ratification or the pre-phase of implementation. Strong or very strong political battles which were observed for the Kyoto-Protocol or for the tropical timber trade regime occurred less frequently than moderate forms of political battles. If strong political battles occurred, they were mainly limited to a small number of member countries. This finding suggests that national governments normally coordinate their positions during negotiations with actors which are possibly affected by future implementation of international policies. The bilateral regimes dealing with fisheries in the Barents Sea or with the management of the Great Lakes have obviously been determined by few political battles on the domestic level during this period. Rather smooth transition from negotiation towards ratification has been identified for the Danube river protection regime, for various elements of the regime developed for preventing intentional oil pollution at seas, or for the Ramsar regime on wetlands. When few or no political battles have been identified as a major characteristic during the period of ratification or implementation, it normally applied to the broad majority of member states. For the majority of regime elements belonging to regimes for the protection of the ozone layer and for the Rhine, few or no political battles characterized the translation of international agreements into domestic obligations, but some elements – notably the Chemical Pollution Convention – caused strong political battles in single countries. Regime experts frequently avoided to submit concrete information on a number of supplementary steps that states have chosen for implementing environmental agreements. Some of these difficulties may have been caused by a too comprehensive list of possible steps that could be checked in our codebook. Information about the steps that were taken by a state in implementing legal agreements is frequently based on information by only one of the two experts coding a regime. A look on single important nations illustrates that a broad range of possible steps were taken for translating international agreements into domestic obligations. Detailed information about the steps taken by the United States exists for ten regimes. More than two dozens of legal agreements (whether they require ratification or other steps) belong to these regimes. For each regime, this country has ratified important legal agreements or taken other steps that are necessary to translate international obligations into domestic practices. In nearly each of these regimes, administrative or non-governmental units were designated to be responsible for implementing the agreement and agencies which were given responsibility in the legislation devised and promulgated more detailed regulations. For nearly each of these regimes data illustrate that responsible agencies initiated programs needed to give effect to the provisions of environmental agreements and that funds were allocated to implement agreements domestically. In several regimes, the United States had already taken various steps before the international agreement was concluded. These previously taken steps facilitated implementation and compliance with rules which emerged on the international level.

Nearly each of the more than 20 international legal agreements for which information about compliance by Russia is available was ratified or acknowledged by this country by taking steps other than ratification. Though Russia or the former Soviet Union have taken a number of supplementary steps to translate these agreements into domestic obligations, it is striking that this country occasionally failed to implement regulations in such a way that behavior could fully comply with international rules. Obviously, this country could by far less rely on already existing policies, administrative responsibilities, or existing social practices on the domestic level which could be used to support implementation in the majority of the nine regimes explored. The data about the steps for implementation taken by Russia or the former Soviet Union on the domestic level suggest that only in a few cases measures were taken or non-governmental units were designated before the agreement was concluded. This country could normally not rely on experiences that were made with the domestic management of environmental problems in other countries. The considerable lack of financial or technological resources and a neglect to collect information about environmentally damaging behavior by domestic actors may further have contributed to arrive at lower levels of compliance in single cases.

Data on a number of industrialized states (e.g., the United Kingdom, Germany, France or Japan) illustrate that some implementing legislation had been passed in these countries already before the international agreement was concluded: non-governmental units were frequently designated, domestic agencies had often devised or promulgated detailed regulations of legislation, domestic agreements were often negotiated with firms or other actors, and funds to implementation measures were occasionally allocated. The existence of financial and technological resources made implementation for industrialized countries less difficult than for other members. These countries had better starting positions than less developed countries for implementation for at least two other reasons: First, they could rely on administrative and legal frameworks or on political support which had emerged when problems were already managed on the domestic level before their trans-boundary relevance was acknowledged by the creation of international regimes. Second, states which are deeply integrated in international society as measured by the number of memberships in international governance systems could collect experience in regard to the measures which have to be taken for effective implementation. But occasional failure by these countries to comply also indicates that such learning must not inevitably lead to desired behavior.

Summary

Full compliance by all regime members is neither the rule nor the exception. Full compliance could be found for the majority of regime elements explored. In single regimes where the majority of regime elements have been determined by compliance of all members, it became obvious that states did not fully conform with norms and rules existing in single regime elements. The frequency distribution on our ordinal scale concerning compliance behavior is not determined by strong differences between cases ending in 1998 or before. However, compared to regime elements which ended before 1998, more demanding norms and rules had to be fulfilled for achieving compliance in some of the regime elements which end in 1998. The expansion of membership which could be observed in many regimes during the 1980s and 1990s could lead us to assume that the emergence of compliance problems becomes more likely. These difficulties especially arise in connection with accession of members which lack capacities to implement the norms and rules of regimes effectively. Observed levels of compliance did not decrease, but environmental regimes were able to cope with the challenges arising for compliance from increasing membership and growing density and specificity of norms and rules. Some regimes have experienced a dynamic towards the achievement of improved levels of compliance as far as behavior of all regime members is concerned. Compliance problems are particularly relevant if important states in the issue area fail to meet regime requirements. Exploring the behavior of important states reveals that a relatively high level of compliance with regime rules can be found for these countries. But problems with compliance by important nations could also be observed in several regimes. In single cases like the ozone regime, developing countries like China or India made provision of financial and technological assistance a condition for their willingness to accede. Finally, it could be shown that states can have an advantage over other states when they can rely on experiences, capacities, or procedures during implementation which they gained partly as a result of integration in international governance systems or which result from high levels of socio-economic development.

Regimes, Programmatic Activities and Compliance

The community of regime researchers frequently put forward the argument that "compliance with international environmental treaties can be ensured at an international level almost exclusively when there is institutional support" (Kiss 1996, 50). A number of questions arises in this context: How did institutions influence compliance behavior? Was compliance achieved by the use of institutional mechanisms which rely on coercion or by mechanisms which intend to manage compliance problems or support capacity-building for implementation on the domestic level? Why does the level of compliance differ among regimes although institutional mechanisms have emerged in all regimes for promoting compliance? Why can we observe ineffective levels of compliance in single cases although regimes provide institutional mechanisms for compliance management?

Table 7.6 correlates the existence of programmatic activities with compliance at the level of all member states. The data indicate that programs for the monitoring or verification of compliance, for the review of implementation or for financial and technology transfer existed in many regimes. Each of the 23 regimes explicitly calls for the conduct of at least one of the three programmatic activities. The broad majority of regimes provide regulations which stipulate the creation of each of these programs. Twenty regimes explicitly provided for compliance monitoring. The provisions of 19 regimes stipulated the creation of mechanisms for the verification of compliance. In each of the 23 regimes, legal provisions demand the establishment of mechanisms for implementation review for at least one regime component. In several regimes, it took a while until these programs became operational. The development and implementation of common standards and of methods aiming to monitor and to verify compliance or implementation review occurred often after a lengthy process. Refinement of these methods and standards takes place more or less regularly, because new technologies or production and consumption patterns can involve new forms of non-compliance. Availability of new technologies offers new opportunities for making the work of single programs more effective. Many regimes experienced increasing relevance of one or various programmatic activities during their lifecycles.

Programmatic activities are normally carried out on the international level. But support by national authorities is necessary for the effective operation of these mechanisms. National authorities collect information and report data to regime bodies or international monitoring networks. The Barents Sea fisheries regime is a case where each of the two member countries monitored compliance by fishermen on the national level. While compliance was controlled at sea and onshore by military coast guards, by civilian fishery inspection services or by domestic agencies, both countries exchanged information about compliance in bodies like the Joint Norwegian-Russian Fishery Commission or the Permanent Committee for fisheries management and control (Honneland 1998, 341). This example illustrates that in single regimes compliance is sometimes less a problem that is caused by states but by private actors which exploit a common pool resource or pollute environmental goods in areas beyond national boundaries. The Co-operative Programme for the Monitoring and Evaluation of the Long-Range Transmission of Pollutants in Europe (EMEP) carried out for monitoring compliance by states with LRTAP-regulations is an example that can be used for illustrating that "there are

Table 7.6   Programmatic Activities and Compliance (Level of All Regime Members)

Total of Regime Elements	Programmatic Activities				At least one of the four programs existed	Compliance by all Member States
	Compliance Monitoring	Review of Implementation	Verification of Compliance	Financial and technology transfer		
18 (13.8%)	10 (13.7%)	10 (16.1%)	8 (12.5%)	3 (7.0%)	10 (55.5 %)	Behavior exceeds regime requirements
62 (47.7%)	26 (35.6%)	24 (38.7%)	30 (46.9%)	15 (34.9%)	41 (66.1%)	Behavior meets regime requirements
34 (26.2%)	27 (37.0%)	16 (25.8%)	21 (32.8%)	19 (44.2%)	28 (82.4%)	Behavior conforms with some requirements but not all
13 (10.0%)	8 (11.0%)	10 (16.1%)	5 (7.8%)	4 (9.3%)	10 (76.9%)	Behavior conforms some of the time and/or to some degree but not completely
3 (2.3 %)	2 (2.7%)	2 (3.2%)	0 (0.0 %)	2 (4.7%)	2 (66.6%)	Behavior does not conform at all
130 100.0%	73 100.0%	62 99.9%	64 100.0%	43 100.1%	91 70.0%	Total

tendencies either to entrust treaty organs with autonomous monitoring powers and the respective infrastructure or to establish integrated monitoring systems with an international evaluation centre" (Marauhn 1996, 708).

Compliance reporting and implementation review are two elements in a more complex causal chain of various activities through which institutions can impact on compliance by states. National self-reporting is necessary for reviewing and assessing state behavior. Reviews are necessary for determining whether further steps must be taken which induce non-complying states to comply. In nearly all regimes, states have an obligation to submit national reports on domestic implementation. Failure to report to international bodies is frequently one among a number of possible features of non-compliance. Measures to improve compliance with national reporting obligations have frequently been implemented in regimes. Reliable information about the existence of reporting procedures which require the submission of information by members about regime implementation could be collected for 22 regimes. In 19 regimes, reporting procedures apply to all components of a regime, whereas in two regimes they only apply to single components. The Black Sea regime was the only institution where explicit reporting procedures were missing.

In the vast majority of regimes various review procedures were formally provided and carried out by the regime. Data indicate that various forms of information gathering about implementation and compliance as well as different types of reviews had become standard practices in international environmental regimes by the end of the twentieth century. Any kind of information gathering about implementation and compliance behavior has been carried out in nearly each regime. At least one of several types of reviews has been carried out at the end of the twentieth century in nearly each regime – whether this review focused on broad assessment of compliance in the regime or on the performance of single members. Other types of procedures for reviewing implementation which intrude far more into the sovereignty of a state have emerged in only a few regimes. On-site inspections to verify compliance are still provided in only a few institutions such like the Antarctic regime, the CITES-regime, in the component of the IATTC-regime which refers to the conservation and management of dolphins, or in the MARPOL-regime. Measures by regime bodies which include recommendations or implementation of responses to inadequate compliance by single members were also provided in only a few regimes at the end of the past century. Such procedures have been used inter alia in the CITES regime, in the ozone and climate change regime, the Ramsar regime on wetlands, or in the MARPOL-regime.

Compliance has mostly been achieved by the use of a management approach. The use of an enforcement approach remained the exception. The management approach has been dominant in about 90 per cent of the 121 regime elements for which data are available. In some cases, it was referred to by experts that exact distinction between enforcement and management is hardly possible. For the two regime elements dealing with developments in the CITES-Convention between 1973 and 1989 and between 1989 and 1998 data by both coders actually reflect disagreement on whether these elements have been determined by enforcement or management, but in a comment one of the two experts indicated that both approaches have been used. For the IATTC-regime it has been indicated that the United States used sanctions in the period between 1980 and 1998 to compel Mexico and other states to implement some controls to reduce dolphin mortality. However, it has been argued that "this 'enforcement' was not the same thing as a sanction put in place by the regime per se" and that through official IATTC channels a management approach has always been applied although management and enforcement have not been mutually exclusive options when outcomes in the issue-area will be assessed.[5]

Compliance mechanisms are steps which can be applied vis-à-vis states which fail to comply. More than half of the regimes explored either completely lacked official compliance mechanisms or provided for them in only a few but not all regime components. Punitive measures like imposition of financial/economic punishments, exclusion of membership or suspension of membership rights were infrequently provided for or used in regimes. Support for capacity-building to enhance compliance or the granting of a transition period to achieve compliance were used more frequently. In some regimes, issuance of notice of violations has been provided for and practiced. States mostly avoided the resolution of disputes about compliance by the use of dispute settlement procedures. In more than half of the 23 regimes, constitutive provisions formally called for the resolution of disputes. But in only a few regimes, these procedures have been used more or less regularly.

Discursive strategies have dominated compliance management, but political pressure complemented this process. On the political scene, measures like threats to punish non-complying states economically were used for putting pressure on these states to comply. From the range of political strategies which can be applied for inducing states to comply shaming has been used most frequently. A few regimes have been determined by the evolution of compliance mechanisms where formal provisions are no longer exclusively dominated by the use of a management approach. The members of the Kyoto-Protocol began to establish a comprehensive compliance system that complements the broad set of mainly cooperative measures agreed upon in the climate change convention with a strong coercive element (Oberthür and Marr 2002). However, this compliance system emerged only after 1998. By the end of the twentieth century, the non-compliance procedure of the Montreal Protocol has been referred to as one of the still rare examples of formal mechanisms in environmental regimes which are identifying

---

5   Written comment to the coding of this variable made by Virginia Walsh. A similar argument concerning the achievement of compliance by the use of unilateral pressure on countries which disregarded decisions of the International Whaling Commission has been made by Sebastian Oberthür (1998, 33). The author argues that the United States threatened to apply sanctions "against non-member whaling states in the 1970s and against the IWC members who continued their whaling activities under an objection after the start of the moratorium".

and handling problems of noncompliance. This procedure can recommend the use of positive or negative incentives for achieving compliance. It is based on regular meetings where members of the Implementation Committee deal with compliance issues. In addition, parties can submit their concern about possible noncompliance by other parties. They can also raise their own compliance problems as a topic on the agenda of the Committee (Victor 1998). In collaboration with international agencies this mechanism managed noncompliance by various East-European countries and accomplished that improved levels of compliance could be achieved.

Which conclusions can be drawn regarding the impact of programmatic activities on compliance? First, the relevance of institutional activities for the management of compliance in regimes increased the more regime rules began to permeate socio-economic practices on the domestic level. Programmatic activities have emerged in all environmental regimes. But compliance behavior was not sufficient in each of the regime elements explored. On the other hand, the level of compliance has not decreased even though during the 1990s regimes required the implementation of measures by states which are often more comprehensive and interfere more strongly in domestic affairs than previous policies. The expansion of programmatic activities for the management of compliance on the international level can partly be understood as a response to the deepening of cooperation in an issue area as "the risk of defection tends to increase the greater, more costly or technically demanding the behavioral change prescribed by the regime" (Underdal 1998, 12). They contribute to increase transparency about state behavior or about implementation and thus also deter states which possibly consider non-compliance.[6] Second, the expansion of programmatic activities in regimes suggests that institutions contributed to achieving improved levels of compliance. Such findings emerge from the coding of regimes with long life cycles. When the life cycles of regimes were divided into different time periods, the impacts of institutional dynamics on compliance behavior which occurred during a period could be compared with earlier developments. In each of these cases, no worsening of compliance behavior could be observed. Rather, satisfactory levels of compliance remained stable although rules became more extensive and demanding. In some regimes, compliance improved by one value on our scale during later periods of a regime. Non-compliance with international catch limits of the whaling regime has determined the behavior of the USSR during the 1950s and 1960s, but "this non-compliance ended suddenly when the international observer scheme was implemented in the early 1970s" (Oberthür 1998, 32). Third, when regimes provide for institutional activities which can be used for reviewing implementation or for monitoring and verifying compliance, it can not be guaranteed that effective levels of compliance will be achieved. Problems regarding the implementation of regime policies occurred in many regimes more or less regularly. But regimes improved the

---

6   See also Thilo Marauhn (1996, 698) who illustrates that compliance control "is a device for generating confidence of states that the benefits of respecting and implementing the obligations of the treaty outweigh the costs".

ability of states to perceive and manage these problems. Programmatic activities ensure that compliance problems will be discovered and can be managed. Their contribution to managing compliance problems helped non-complying states to increase their capacities for implementation on the domestic level. An expansion of activities for the management of compliance was observed in regimes. Fourth, the institutional design is obviously an important condition for achieving compliance. But variance in compliance behavior can not sufficiently be explained by the impact of institutional mechanisms alone. Other factors like the willingness of states to take requirements for reporting or implementation of international policies seriously also determine whether effective levels of compliance can be reached. Unfavorable conditions for achieving compliance can emerge from the specific problem type that determines an issue-area. The number or type of actors whose behavior must be controlled or the type of environmental good managed by a regime are only a few of a number of possible factors which intervene and must be considered by institutional mechanisms in managing compliance.

The Impact of Non-State Actors on Compliance

States acknowledge obligations in international treaties by accession, ratification or by other steps. Non-compliance can mean for a state that its reputation in international society will be damaged. But the possible danger of a loss of reputation does not always motivate states to comply. Resistance from domestic actors or state failure are frequently factors which prevent that states comply with international norms. Non-state actors can be causers of compliance problems. But they can also contribute to improving compliance. How significant is the role of non-state actors for compliance by states? How can these actors contribute to managing compliance in international regimes?

In the following, two levels of analysis will be distinguished in exploring the impact of these actors on compliance with international norms and rules. First, the impact of non-state actors will be illustrated on compliance behavior of states. In connection with some of the variables which measure aspects related to compliance behavior by states, the impact of non-state actors and of governmental policies on these outcomes has been assessed by experts during the coding. Non-state actors will be explored in terms of their roles as causers of compliance problems as well as in regard to their contribution to implementation and compliance. Second, attention will be paid to participation of non-state actors in different regime functions. They deal with the monitoring or verification of compliance or with the assessment and review of state performance. Some of these functions are carried out by non-state actors also on the domestic level where they support governments in the collection of information or in implementing international policies.

Which findings emerge from data concerning the impact of non-state actors on compliance behavior by states? First, coding experts indicated for each of the 23 regimes that government policies were those events and actions that were significant in the regime's causal impact on compliance. On the domestic level, governmental activities were normally dominating forces for producing compliance. This finding is not a surprise, since governments have main responsibility for implementing international policies. Non-state actors were also ascribed a causal role for these outcomes in the majority of regimes explored. These findings lead to the conclusion that (in combination with governments) non-state actors were relevant for causing compliance by states. One has to admit that various forms of governmental policies have been referred to more frequently as types and events that played a significant role in the regime's causal impact than activities of non-state actors. Second, watchdog groups pushing for regime compliance have been referred to far more frequently as causal factors for compliance than changes in industrial goals or methods not required by government policies. Structural economic change played a significant causal role for compliance in cases like the Great Lakes management regime, the river regime for the protection of the Rhine, or the ozone regime. While in each of these cases watchdog groups were ascribed similar relevance for compliance than change in industrial goals or production methods, they have been referred to also for a number of other regimes where structural economic change was considered less relevant. The ability of the economy for developing environmentally sound technologies and production methods has certainly grown. But these findings also suggest that societal pressure is often necessary for pushing governments and the economy to take those steps which are required that a state will comply with international regulations. Third, there has been some variance among the regimes explored concerning the causal relevance ascribed to non-state actors for implementation and compliance by states. A combination between the causal relevance of government policies and activities of non-state actors for compliance can be derived from the data that were collected for single elements of the following regimes: regimes for the protection of the North Sea, for the Great Lakes, the London Convention regime, the Antarctic regime, the Baltic Sea regime, the CITES regime, the whaling regime, the Ramsar regime on wetlands, or the regimes for the protection of rivers like the Danube or the Rhine. For a few other cases like the climate change regime only government policies have been identified as causally relevant, but no significantly lower levels of compliance were detected in this regime until 1998.

The positive and negative impacts that arise from an international agreement for social groups can influence implementation on the domestic level and compliance by a state. Coding experts were confronted with the difficulty to assess the impacts of international regimes on social groups in various important nations. For some elements, this assessment could not be made. Coding Experts indicated that they did not have the detailed knowledge of individual countries needed to provide accurate information. These problems are understandable. They may cause the community of regime researchers to spend more efforts on exploring the impacts of international norms on domestic and transnational levels. For some regimes, no data could be provided concerning the impact of a regime on social groups for other reasons. For example, the narrow functional scope and the shallow character of the rules of the Vienna Convention as the founding document of the ozone regime involved that "no significant positive or negative effects on social groups can be detected" for this convention.[7] As far as the desertification convention is concerned, it has been argued that because "the convention addresses such a broad issue as sustainable development in areas experiencing land degradation (plus that it has not been in force for very long), it is impossible to make an assessment of whether behavior has been affected".[8] Nevertheless, data about the positive and negative impacts of regimes on social groups could be collected for the majority of cases. These data make it possible to understand whether compliance has been made more difficult or facilitated by the impacts of a regime on social groups. Distinction between positively and negatively affected social groups can be artificial in single cases. For the ozone regime, it has been argued by a coding expert that some of the social groups were affected both positively and negatively. While implementation of the legal agreements which constitute the ozone regime "required some adaptation and caused related costs (in particular additional investments)", such adaptation and change "frequently caused additional benefits, e.g., significant process and production innovation or lower operational costs".[9]

In a first step carried out during the coding those social groups were identified which where positively or negatively affected by the operation of a regime on the level of important nations. In a second step, it was determined to what degree social groups where affected in important nations. To this end, two independent scales have been used for determining the degree to which a social group in an important state was affected positively or negatively. Major positive effects and little or no positive effects were used as upper and lower limits on a scale from one to three for measuring positive impacts. Major positive effects subsumed dramatic developments such like a strong increase in economic turnover or significant improvements of the situation occurring for those people which were affected by an environmental problem. Little or no positive effects applied to a situation where the benefits received by a group were only marginal. Likewise, major negative effects and little or no negative effects were used as upper and lower limits on a scale from one to three for measuring the degree to which social groups were negatively affected. For example, major negative effects existed for a group if a dramatic increase of costs (e.g., significant decrease in economic turnover, significant costs for investments in new technologies) or other dramatic behavioral changes (e.g., increase in gasoline prices for car drivers by more than 20 per cent) occurred. Little or no negative effects were identified if adjustments which had to be made by a group were only marginal.

For 15 regimes data are available about positively or negatively affected social groups. For less than half of these regimes, major positive or major negative effects have been identified for a social group in one or more important states. Major negative effects for social groups were just as infrequently identified as major

---

7   Written comment to this variable made by Sebastian Oberthür.
8   Written comment to this variable made by Elisabeth Corell.
9   Written comment to this variable made by Sebastian Oberthür.

positive effects. Of the more than 400 data that exist about negatively affected groups in important states less than 15 per cent apply to major negative effects. More than half of these data refer to marginal negative effects. About a third of these data apply to moderate negative effects. Of the more than 600 data that exist about positively affected groups, about half refer to marginal positive effects. Nearly the same amount of data indicates moderate positive effects for social groups. Only a bit more than 5 per cent of the data refer to groups which experienced major positive effects. These findings suggest the conclusion that nothing impossible has been required from social actors by policies for implementation in the broad majority of cases. Major behavioral changes can be required from social groups for achieving compliance. But such negative effects – whether they involve economic costs or other burden – did normally not cause serious levels of non-compliance in the vast majority of cases.

Nevertheless, major negative effects arising for social groups in single countries from international policies for the conservation of natural resources or for the reduction of environmental pollution can involve that domestic resistance will prevent compliance by a state. Different observations can be made regarding the impact of negatively affected social groups on compliance. A few examples exist where major negative effects for social groups correlate with non-compliance by a state. But many other examples also illustrate that major negative effects for social groups must not inevitably lead to low levels of compliance by these states. The whaling industries of Norway and Japan have suffered major negative effects from regulations of the whaling regime. But behavior of both countries mostly met regime requirements. A large causal influence has been ascribed to the whaling regime for the behavior of both countries.[10] In the majority of important states a regime produced positive as well as negative effects for single groups. Lobbying of negatively affected groups against implementation of international policies has often been countered by beneficiaries of these policies. Positive benefits were not only measured for economic actors or for people affected by environmental problems. Activist environmental non-state actors or scientific and technical service organizations were frequently identified as groups which benefited from regime policies. Such positive effects emerged when these actors became more involved in domestic implementation or were ascribed growing relevance in domestic policymaking.

The antagonism between negatively and positively affected groups must not be permanent. Impacts of regime policies on the behavior of social groups can also change during the life-time of a regime. For example, it has been indicated for

---

10   Jennifer L. Bailey (2008) comes to a more critical assessment of the developments which took place in the whaling regime after the moratorium on the commercial take of all species of great whales had been adopted in 1982. She argues that efforts to establish an anti-commercial whaling norm in the regime failed. Many key states have never accepted a protectionist norm. States like Norway, Japan, or Iceland resumed commercial whaling in the 1990s or later on.

the IATTC-regime that "over the short term the industry suffers because of limits on production as a result of the conservation program, but over the long term the industry benefits from increased tuna populations as a result of the conservation program".[11] A similar observation has been made for the ICCAT-regime where a coding expert indicated that "the imposition of management measures means cut backs in the operations and catch causing economic and social losses, but these same measures mean that fishing and processing opportunities are maintained in later years, or enhanced: both giving economic and social benefits".[12]

Which roles can non-state actors play in managing compliance on the level beyond the state? In many regimes, watchdog-groups and service-oriented non-state actors made efforts that states were implementing international norms and rules. Politicization of non-compliance is one among various important contributions through which these actors can impact on compliance by states. Detailed information about such activities of non-state actors in regard to compliance has not been collected during the coding. But a long list of cases could be referred to which can illustrate that awareness raised by non-state actors for non-compliance of single states remarked the beginning for improving compliance in these states in the long-term.[13] The contribution of these actors to compliance management has not been addressed systematically during the coding. But empirical information collected in connection with the coding can be used for illustrating the growing relevance of non-state actors for compliance management. This contribution is largely made in the context of institutional mechanisms such like monitoring and verification of compliance or implementation review. It can not be clearly separated from the causal influence of a regime on compliance, but it is often more or less part of the operation of a regime. In some respects, these activities are often accompanied by activities of governments or governmental agencies. Ronald B. Mitchell (1994, 297–8) concluded for the regime for the management of international oil pollution at sea that the changes in behavior and variance on compliance levels can be attributed "to the success of treaty provisions in accomplishing three tasks: creating 'opportunistic' primary rule systems that impose requirements on those actors most likely to fulfill them, creating compliance information systems that give information providers a 'return on their investments', and creating noncompliance response systems that remove international legal barriers from those actors with incentives to respond to noncompliance". Nevertheless, the author also points to the fact that non-governmental actors have made important contributions to managing compliance.

From the set of regimes included in the IRD, the CITES-regime offers a very interesting example in this regard. The wildlife trade monitoring network TRAFFIC

---

11   Written comment to this variable made by James Joseph.

12   Written comment to this variable made by James S. Beckett.

13   Examples for regimes where activities of non-state caused improvements in compliance behavior by single states are provided in Chayes and Handler Chayes (1995, 259–69).

is a joint program of IUCN and WWF which has been established more than two decades ago for the monitoring and documentation of international trade with endangered species. Its findings derived from monitoring and research of wildlife trade became established in the work of CITES and made a significant contribution for improving compliance with the norms and rules of this regime. WWF has taken the role of a lead party for implementing management plans for coastal lagoons and wetlands in the Baltic Sea area which represents one of the six program elements of the implementation of the Baltic Sea Joint Comprehensive Environmental Action Programme (JCP) (Helcom 1999, 30–31). The tasks that must be fulfilled in connection with the verification and management of compliance or concerning the production and review of reports about national implementation have increased in regimes. Growing involvement of non-state actors has not taken place at the expense of the role of the state. Non-state actors could contribute to these regime functions because they could not be carried out effectively by the state alone.

The question remains on whether correlation can be drawn between growing involvement of non-state actors in compliance management and observed levels of compliance. Experiences made with the above-mentioned regimes seem to suggest that such an impact of non-state actors on compliance can be identified. But this contribution can often not be considered independently of the causal impact of a regime. It must also be considered that implementation and compliance are mainly caused by government policies. But failure or success of these policies is often also dependent on participation or political support of non-governmental actors. Non-state actors considered international institution-building as an opportunity to contribute their services for implementation and the management of compliance. But they could only provide these services because regimes had established the programmatic activities which can be used for implementing regime policies and for managing compliance problems. States integrated non-state actors in the work of regimes. But states mostly kept the dominant role in the management of compliance. Growing depth of global environmental governance can possibly foster the trend towards further inclusion of non-state actors in the work of compliance mechanisms.

Conclusion

For the majority of regimes effective levels of compliance or over-compliance could be identified. In some regimes with watersheds which separate regime components into different time periods, improvements in compliance occurred on the level of all members or in single countries. Levels of compliance did normally not worsen although the functional scope and the density and specificity of rules broadened in many regimes over time. This finding coincides with the observed expansion of institutional mechanisms in regimes which review implementation or which monitor or verify compliance. In providing regimes with these mechanisms, states have begun to improve the capacity of international institutions for producing compliance. Regimes promoted compliance as one among various grounds for legitimacy. Even though nearly each regime provides at least one of these institutional mechanisms, insufficient levels of compliance have been identified for a number of regimes. This finding suggests that the stimulus coming from these institutional mechanisms is in some regimes still not powerful enough for producing effective levels of compliance. The growing depth of cooperation can involve that the use of a mainly discursive procedure will no longer be sufficient for achieving compliance in the long-term. In some regimes compliance mechanisms must possibly be complemented by court-like procedures and punitive measures. The existence and effective operation of compliance mechanisms alone can not always guarantee that cases of non-compliance can be avoided. But programmatic activities can increase transparency about state behavior, reveal problems occurring in connection with implementation on the domestic level, or detect deliberate cheating. Compliance mechanisms are causally relevant for achieving compliance. But one should not be blind to the fact that compliance can also depend on other factors like political backing for implementation or the availability of capacities. For the vast majority of regimes, non-state actors made significant contributions during implementation and compliance management. On the domestic level, governments must partly rely on the expertise of non-state actors during implementation or can secure the adoption of implementing policies only by the political support mobilized by watchdog groups. On the international level, the information relevant for assessing whether or not states comply would be less complete if non-state actors would not contribute their services or politicize violations against international norms and rules. Institutional mechanisms dealing with implementation review or with verification and monitoring of compliance can often fulfill their tasks only in a sufficient way if they rely on the expertise of governments or non-state actors. Attention has been directed on the discursive power of these actors in processes of domestic or international politics. In the transnational political public, politicization of non-compliance of a state by non-state actors can be important for achieving improved levels of compliance. It became also obvious that non-state actors are often involved in complex policy-networks in regimes dealing with programs for the implementation of regime policies or the management of compliance. Nevertheless, variances in the levels of compliance do not inevitably always correlate with variances regarding the participation of non-state actors in implementation or compliance management. Issues related to implementation and compliance are still frequently considered as a domain of governments in regimes. Participation of non-state actors in compliance management or their political roles both reveal that non-state actors can contribute to improving the implementation of international policies or compliance in regimes. These findings support our prime hypothesis. But one has to admit that many of the contributions made by non-state actors could not take effect without the existence of institutional mechanisms. Institutional mechanisms must exist which can be used for dealing with the causes of this form of state behavior and for developing solutions which can improve compliance behavior.

Chapter 8
Regimes and the Management
of Environmental Problems

Social rule deserves obedience if it can contribute to the well-being of a community.[1] The obligation of social rule to increase the utility of the community to which it is accountable finds expression in the development and implementation of international policies. Two different standards shall be applied for assessing whether regime policies contributed to the effective management of environmental problems: i) the study of *goal-attainment* focuses on exploring whether the goals established in international regimes have been achieved; ii) the study of *problem-solving* intends to assess the changes in the state of the problems managed by regimes. First, it will be measured whether goals have been fulfilled and whether the state of the problems has changed. Subsequently, it will be explored whether regimes and non-state actors contributed to the fulfillment of these goals and to changes in the state of problems. Analysis starts from the assumption that goal-attainment and problem-solving are the result of a complex causal chain. It can be expected that considerable improvements in the cognitive setting and in compliance behavior are preconditions for the attainment of environmental goals or for achieving considerable improvements in the state of environmental problems. In connection, it can be expected that impacts of institutions and of non-state actors on the cognitive setting or on compliance could have similar effects.

Goal-Attainment and Problem-Solving

Did regime members accomplish the important goals which they agreed upon in regimes? Did the state of a problem change during the period of a regime element? Which causal impact was ascribed to regimes for these developments? In assessing the causal impact of regimes on goal-attainment and on the state of the environment, a stock-taking will be made which informs on whether the significant efforts that could be observed in global environmental politics have borne fruit in the past few decades. The frequency distribution for data on the different scales

---

1   The literature on international relations is full of evidences that the desire to improve the outcome of policies strongly motivated the historical evolution of international governance systems. These expectations arose from the assumption that independent state action leads to outcomes that are less optimal than those produced by international collaborative action (Jacobsen 1984; Rittberger and Zangl 2003, 159f).

used for determining whether goals have been fulfilled or for measuring changes in the state of a problem will be illustrated for all regime elements and for a sub-set of regime elements ending in 1998. The intercoder reliability will be checked for those data which refer to regime elements coded by two experts independently.

Regimes and the Fulfillment of Goals

With the previous chapter dealing with uncertainties in the cognitive setting it became obvious that more than one problem can exist on the level of a regime element. Likewise, more than one goal can normally be found in a regime element. How are goals and problems connected with one another? The goals which are established by states in international treaties are often less complex than the problems which exist in issue-areas. The various goals formulated by states in a treaty normally consist of targets that are more complex than regime rules. In ideal circumstances, various goals of a regime are interrelated and their fulfillment will lead to improving the state of a problem. Functional goals which aim to improve transnational cooperation in scientific research, monitoring, in the exchange of data or in the development of policies must be achieved before environmental goals can be realized. Environmental goals can involve targets which correspond to or which are less far-reaching than the ideal solution represented by the collective optimum. It has been ruled out for methodological reasons to use the collective optimum as a benchmark for assessing the degree of problem-solving in environmental issue-areas. Even so in some regimes the goals established by states demand nothing less than the achievement of the collective optimum. When studying goal-attainment, the question must be raised whether the collective optimum has been achieved or not. In any case only a small minority of goals explicitly calls for the collective optimum as a target.

The important goals of regimes were identified together with coding experts before the actual coding began. The coding experts of a regime reached agreement about the character and wordings of these goals. For example, Article 2 of the 1979 Convention on Long-Range Air Pollution defines as a common goal that "parties endeavor to limit and, as far as possible, gradually reduce and prevent air pollution, including long-range trans-boundary air pollution" (United Nations 1996, 3).[2] Important goals that are included in a constitutive agreement like a framework convention frequently pertained to several regime elements. In this case, a goal has been coded for each individual regime element separately. For some regimes coding experts identified only one goal for a regime element. But for other regime elements, several goals were considered important. Accordingly, it would be inadequate to only assess how many of these goals have been fulfilled or remained unrealized. It will also be explored whether goals have been fulfilled on the level of a regime element or for a regime as a whole.

---

2   This environmental goal pointed the way ahead for additional international policies that were developed in subsequent legal protocols in the 1980s and 1990s (Levy 1995; Wettestad 1999 and 2002a).

A binary scale has been used for exploring goal-attainment. It distinguishes whether goals have been fulfilled or not during the lifetime of a regime element. Data about goal-attainment exist for a total of 524 goals (see Table 8.1). This data-set includes 150 of the total of 172 regime elements. It covers all 23 regimes. Four hundred and one of these data (76.5 per cent) indicate that goals have been fulfilled. Hundred and twenty-three goals (23.5 per cent) were not fulfilled. No other finding emerged for the goals belonging to the subset of regime elements which end in 1998. Two hundred and thirty-two pairs of goals (involving a total of 464 goals) can be checked for inter-coder reliability. Agreement between coding experts existed for 176 pairs or 352 goals (75.9 per cent). These pairs of goals belong to 120 regime elements from 16 regimes. Three hundred and twelve goals (88.6 per cent) of this particularly reliable data-set were fulfilled. Forty of these goals were not fulfilled (11.4 per cent). The data indicate that the broad majority of important goals that were established by states in international environmental regimes have been fulfilled. The above-described differences concerning the number of goals that were identified for single regimes lead us to raise the question on whether a different finding will emerge if analyses will not focus on single goals. Instead, the focus can also be directed on regime elements or on a regime as a whole. In the broad majority of regimes, the number of attained goals exceeded the amount of goals which remained unfulfilled. Failure to achieve single goals occurred in the majority of regimes. But in most of the regime elements this failure coincided with the fulfillment of other goals. In a few regimes, such failure has been widespread or determined developments in regime elements – in particular when coding experts primarily focused on the coding of comprehensive environmental goals. Occasionally, environmental goals which called for the fulfillment of the collective optimum remained unfulfilled by the end of the twentieth century.

A few examples can be used to illustrate these findings. The goal of the conservation of whale stocks was achieved during the second period of the whaling

Table 8.1   Goal-attainment

All goals		176 pairs of goals (complete agreement in the codings)*		114 pairs of goals which end in 1998 (complete agreement in the codings)*		Goal-attainment
123	(23.5 %)	40	(11.4 %)	26	(11.4 %)	No (Goal not fulfilled)
401	(76.5 %)	312	(88.6 %)	202	(88.6 %)	Yes (Goal fulfilled)
524	**100.0 %**	**352**	**100.0%**	**232**	**100.0%**	**Total**

* Two codes exist for each goal in a regime element. Pairs of goals were omitted, if no agreement existed between coding experts. Accordingly, the subset of 176 pairs of goals ignores 56 pairs (or 112 data). The subset of 114 pairs of goals coded for regime elements ending in 1998 ignores 42 pairs of goals (or 84 data). Percentage figures were rounded up and down.

regime between 1982 and 1998. But the regime failed to achieve this goal during an earlier period from 1946 onward. The Barents Sea fisheries regime and two other regimes that were established for the conservation and management of tunas or tuna-like fish in the Eastern Pacific or in the Atlantic achieved goals which involved maintaining fish stocks, harmonizing the conservation and economic use of fish resources, or protecting single species. The goal established in the 1963 Convention on the Rhine against Pollution, which stipulates protection of the Rhine against pollution and prevention of further pollution, has been fulfilled just as much as the goals included in the Rhine Action Plan of 1987 or in the two conventions for the protection of the Rhine against chemical and chloride pollution (Bernauer 1996). Some of the environmental goals which emerged for the Great Lakes management regime have been accomplished. For example, the United States and Canada achieved to reverse eutrophication of Lakes Erie and Lake Ontario and to maintain oligotrophic conditions in Lakes Huron and Superior through reduced loadings of phosphorus and other nutrients.

The broad majority of the goals that exist for various components of the Antarctic Treaty regime have been fulfilled. Goals like the reservation of Antarctica to peaceful activity or the avoidance of international conflict about Antarctic territory can predominantly be perceived of as security goals. But only the fulfillment of these goals guaranteed that Antarctica largely remained an unspoilt territory that could be protected from environmental damages which would have probably resulted from the deployment and operation of military facilities. These measures have been complemented by other activities (e.g., a ban on mineral and hydrocarbon exploitation by the 1991 Protocol on Environmental Protection). Likewise, the regime component managing the conservation of seals in the Antarctic region accomplished its environmental goal which focuses on the protection of Antarctic seal populations by limiting or banning seal taking at sea (Joyner 1998; Peterson 1988; Stokke and Vidas 1996). The Antarctic Treaty system is also a good example for illustrating that functional goals were relevant for the attainment of environmental goals. The goal to promote international scientific cooperation in Antarctica provided in the Antarctic Treaty can be considered as an element that becomes important in a more complex chain of steps that must be taken for achieving environmental goals.

Functional goals established in the Barents Sea fisheries regime which provided that Norway and Russia were urged to coordinate marine research, strengthen good neighborliness, or develop co-operation of the fisheries sector in both countries by exchange of information and experience have been achieved. The regime's environmental goal to carry out measures which can secure conservation, reproduction, or sustainable and rational use of living marine resources has also been attained. Nearly all major environmental and functional goals which were established in the various components of the London Convention regime could be fulfilled. Several functional goals had to be fulfilled in the London Convention regime for accomplishing the environmental goal of the prevention of marine pollution caused by the dumping of various sorts of wastes.[3] Various functional or normative goals were established during the lifetime of the ozone regime. Cooperation in science and technological research, observation and monitoring was considered as a major goal by states in the 1985 Vienna Convention for the Protection of the Ozone Layer. But only the existence of other functional goals made it possible that the environmental goal will be achieved in a long-term perspective (Benedick 1991; Parson and Greene 1995).

On the one hand these data produce the encouraging finding that many of the goals that were fulfilled were targeting on improvements in the state of the environment or in the conservation of natural resources. On the other hand, the majority of goals which could not be attained consist of environmental goals. For example, environmental goals established in global regimes dealing with climate change, desertification or biodiversity remained largely unfulfilled by the end of the twentieth century. A similar situation existed in the hazardous waste regime where many institutional components were created in the 1990s only. Environmental goals which were established by OECD-countries or the European Union or which were included in Lome IV-regulations were fulfilled. But some goals included in the Bamako and Waigani Conventions or in the Basel Convention remained unfulfilled.

Variables like the complexity of a goal or time have influenced the likelihood of goal-attainment. In each of these regimes the management of problems gained momentum only in the 1990s and the implementation of policies has often been in its initial stages. Some of the environmental goals that were established in single regimes can hardly be attained by one regime alone. In many cases, goals can often be attained if additional policies will be developed in neighboring issue-areas.

Which causal impact did regimes have on goal-attainment? Two analytical steps had to be taken by coding experts in making causal judgments about the impact of a regime on goal-attainment. First of all, a counterfactual analysis had to be made which considered which developments would have occurred in the absence of a regime. If goals would have been attained without the regime, the regime had no causal impact. A positive causal impact has been ascribed to a regime if it led to some or significant behavioral changes which would have been less likely in the absence of the regime. It had a negative causal impact if behavioral changes would have been more far-reaching in the absence of the regime. Secondly, regime factors that stem from a regime's existence were separated from non-regime factors which operated outside a regime's environment but influenced developments in the issue-area (e.g., economic recession). Four ranks were distinguished on the scale

---

3 In particular, assessments had to be made for exploring the practical availability of alternative land-based options of treatment, disposal or elimination of wastes. Access to, and transfer of, environmental sound technologies particularly to developing countries was facilitated to promote modification of production processes or recycling. Monitoring activities were set up and carried out to guarantee that the environmental goals of this regime could be accomplished.

that was used for determining the causal impact of a regime on goal-attainment. Various dimensions of the positive causal impact of a regime were considered by three values of our ranking which used little or no causal impact and large causal influence as lower and upper limits. Little or no causal impact of a regime occurred when non-regime factors mainly accounted for all behavioral changes and regime factors were insignificant, or if counterfactual reasoning led experts to conclude that similar developments would have occurred also in the absence of a regime. A large causal influence could be found when the regime accounted equally with non-regime factors for behavioral changes or has proven to be more important in this regard than non-regime factors. In addition, this form of regime impact was identified if observed developments would not have occurred in the absence of a regime. The scale took also into account that a regime could have a negative causal influence.

Findings included in Table 8.2 indicate that for the vast majority of the goals which were fulfilled the regime in question has been ascribed at least a modest causal impact. But data also indicate that a large causal impact has been identified more frequently than a modest causal impact. A large causal impact for the attainment of goals was identified for a broad number of regimes. The following regimes were among the cases where both coding teams agreed on the attainment of goals and about the large causal impact of regimes: the Antarctic regime, the ozone regime, regimes for the protection of the Rhine and the Danube, the London Convention regime, the ICCAT-regime, and the Great Lakes regime. In many other regimes, a mix between a modest or large causal impact for the attainment of goals prevailed. Regimes were not held responsible for the non-attainment of goals.

Problem-Solving

The approach which focuses the measurement of relative changes in the state of a problem in a regime element uses a less demanding standard than the collective optimum. Various reasons were given in a previous chapter (see Chapter 3) which illustrate that the methodological pitfalls that exist in connection with the identification of the collective optimum speak against the use of this approach in a comprehensive coding project. The character and wording of a problem were identified together with experts before the coding. The broad majority of these problems described environmental issues. For example, the regime for the protection of the Rhine managed the problem of "pollution causing damage to ecosystems and water quality especially in downstream countries".

In other regimes, environmental, economic or other political issues were either part of one and the same problem or were treated as independent problems. For example, the problem of the South Pacific fisheries regime involved "coordination of fisheries management among the members of the South Pacific Forum in order to i) regulate tuna harvest by distant water fishing nations and ii) maximize returns to the Pacific Island Countries and Territories". In some regimes, environmental and economic problems were disentangled because they were treated as distinguishable

Table 8.2 Causal Impact of Regimes on Goal-Attainment

Goal-attainment	All Goals		Causal Impact of Regimes				
			Not applicable	Little or no causal impact	Modest causal influence	Large causal influence	Negative causal influence
Goal not fulfilled	116	(22.4%)	19 (100.0%)	38 (63.3%)	36 (26.1%)	22 (7.3%)	1 (100.0%)
Goal fulfilled	402	(77.6%)	0 (0.0%)	22 (36.7%)	102 (73.9%)	278 (92.7%)	0 (0.0%)
Total	**518**		**19** (100.0%)	**60** (100.0%)	**138** (100.0%)	**300** (100.0%)	**1** (100.0%)

issues in a regime. Precoding negotiations for the tropical timber trade regime led to the identification of two separate problems which distinguished between environmental and economic issues. This regime dealt with the environmental problem of "increased evidence of significant levels of tropical deforestation and rainforest degradation" in developing countries. It also managed the economic problem of the "underdevelopment of a commercially viable tropical timber industry". In other regimes, environmental and economic issues were considered as parts of one and the same problem because regime members avoided separating them or because environmental issues were given priority by political management. The changes in the state of a problem that occurred during the time period of a regime element were measured on an ordinal scale from one to five. Considerable improvement or considerable worsening were defined as upper and lower limits on that scale. The middle of that scale pertained to continued existence of the status quo. The rank of a slight worsening of the problem took those developments into account which were less dramatic than a considerable worsening but which gave more cause for concern than continued existence of the status quo. The rank of slight improvements considered developments that were more favorable than the former status quo but which did not amount to the level of considerable improvements.

For 195 problems data are available which illustrate changes in the state of the world during the periods which delimit regime elements (see Table 8.3). These problems belong to 147 regime elements and cover all 23 regimes coded. The improvements were slightly (51) or considerable (50) for more than the half of these problems. Positive developments were observed for 51.8 per cent of the

Table 8.3   Problem-Solving in Regime Elements

All problems		57 pairs of problems (codings at most differing by one value)*		32 pairs of problems which end in 1998 (codings at most differing by one value)*		Problem-Solving
19	(9.7%)	13	(11.4%)	1	(1.6%)	Problem worsened considerably
36	(18.5%)	19	(16.7%)	7	(10.9%)	Problem worsened slightly
39	(20.0%)	22	(19.3%)	14	(21.9%)	Problem stayed the same
51	(26.2%)	33	(28.9%)	15	(23.4%)	Problem improved slightly
50	(25.6 %)	27	(23.7%)	27	(42.2%)	Problem improved considerably
195	**100.0%**	**114**	**100.0%**	**64**	**100.0%**	**Total**

* Two codes exist for each problem in a regime element. Pairs of problems were omitted, if disagreement between coding experts has reached more than one value on the scale. Accordingly, the subset of 57 pairs of problems ignores 14 pairs (or 28 data). The subset of 32 pairs applying to problems in regime elements which end in 1998 ignores 8 pairs (or 16 data). Percentage figures were rounded up and down.

problems coded. The state of the world stayed the same for 39 problems (20 per cent). A slight worsening (35) or a considerable worsening in the state of the world has been observed (20) for 28.2 per cent of the problems. A friendlier picture emerges if only those regime elements will be analyzed which end in 1998. A view on these regime elements reduces the statistical distortion that exists in our total data set. This distortion results from the overrepresentation of problems in regimes which have one or more watersheds. One hundred and ten problems have been coded for a subset of 90 regime elements which end in 1998. These regime elements cover all 23 regimes. Slight (26) or considerable (46) improvements developed for 61.8 per cent of the 110 problems coded from this subset. The data-set which can be checked for the intercoder-reliability of those data where two independent codings exist for each problem in a regime element gives a similar impression (see Table 8.3). From this subset of 114 data (or 57 pairs) on problems, 60 (52.6 per cent) indicate either slight or significant improvements in the state of the world. Thirty-two data (28.1 per cent) refer to slight or considerable worsening of a problem. A smaller subset exists for 32 pairs of problems which end in 1998 and where codings differ by no more than one value on our scale. From the 64 data which are part of this subset, 42 (65.6 per cent) are distributed over slight or considerable improvements. The problem remained unchanged or it worsened until the end of 1998 in approximately a third of the instances. All in all, these figures lead to the conclusion that slight or considerable improvements in the state of the problems have been achieved in many regimes by the end of the twentieth century. There are still considerable differences between single regimes concerning the degree to which changes in the state of problems can be observed. For a number of problems developments were less encouraging or sometimes even negative. But data also revealed that negative developments were halted in some regimes or turned towards a desirable direction.

What findings can be gained from assessments for regimes as a whole? The findings included in Table 8.4 suggest that three groups of regimes can be distinguished. These groups differ with respect to the trends that could be observed in the state of problems for the regime as a whole until the end of the twentieth century. A first group of 16 regimes has been determined by *slight or considerable improvements* until the end of 1998. In some regimes of this group, improvements were more obvious than in others. In 6 of these 16 regimes considerable improvements could be observed for a regime in the aggregate. The sub-group of regimes where considerable improvements occurred consists of the LRTAP-regime, the Rhine-regime, the CITES-regime, the fisheries regime for the Barents Sea, the London Convention regime, and the regime for the prevention of intentional oil pollution at sea. The other sub-group where overall developments were determined by slight improvements includes the Antarctic regime, the Danube regime, the regimes for the protection of the Black Sea and the Baltic Sea, the Great Lakes Management regime, the hazardous waste regime, the whaling regime, the ICCAT-regime, the IATTC-regime, and the South Pacific fisheries regime. For a second group of 5 regimes the state of problem-solving that has emerged for the

Table 8.4   State of Problem-Solving for Regimes as a Whole*

Regimes	Total of problems	Problem worsened considerably	Problem worsened slightly	Problem stayed the same	Problem improved slightly	Problem improved considerably
Antarctic Regime	22	0	3	7	5	7
Baltic Sea Regime	8	0	0	4	1	3
Barents Sea Fisheries Regime	2	0	0	0	0	2
Biodiversity Regime	2	1	1	0	0	0
CITES	6	0	0	0	1	5
Climate Change Regime	5	0	5	0	0	0
Danube River Protection	2	0	0	0	2	0
Desertification Regime	2	0	0	2	0	0
Great Lakes Management Regime	3	0	1	1	0	1
Hazardous Waste Regime	3	0	0	0	3	0
IATTC Regime	4	1	0	0	0	3
ICCAT Regime	2	0	0	0	2	0
International Regulation of Whaling	4	1	0	1	0	2
London Convention Regime	6	0	0	0	0	6
Long-Range Transboundary Air Pollution	5	0	0	0	2	3
North Sea Regime	4	0	2	0	2	0
Oil Pollution Regime	2	0	0	0	0	2
Protection of the Rhine Against Pollution	8	0	0	0	0	8
Ramsar Regime	2	0	1	1	0	0
Regime for Protection of the Black Sea	4	0	0	0	4	0
South Pacific Fisheries Forum Agency Regime	2	0	0	0	2	0
Stratospheric Ozone Regime	8	0	1	7	0	0
Tropical Timber Trade Regime	4	1	0	1	2	0
Total	**110**	**4**	**14**	**24**	**26**	**42**

* Data are pertaining to problems in regime elements which end in 1998.

regime as a whole until the end of the twentieth century can be characterized by *continued existence of the status quo*. Some of the regimes that are included in this group were characterized by reverse developments where improvements in single problems coincided with a slight worsening of other problems. The tropical timber trade regime, for example, was determined by reverse developments. The economic problem of the underdevelopment of a commercially viable tropical timber industry improved slightly in this regime. But no progress was made regarding the regime's environmental problem of increased evidence for significant levels of tropical deforestation and rainforest degradation in developing countries. One of the two coding experts made the very skeptical assessment that a considerable worsening of this problem could be observed between 1983 and 1998. One could also argue from an environmentalist perspective that this regime should be included in the group of regimes where overall developments worsened. Four other cases belong to this group. Improvements in the environmental situation of the North Sea were less far-reaching than in other regional seas. Data indicate for the Ramsar regime on wetlands or the desertification regime that at best no further worsening of the problem occurred. A special case in this group of regimes is the ozone regime. The trend of a growing reduction of stratospheric ozone could be stopped and no further worsening occurred. One could argue that such a view considers developments in the issue-area too optimistic because it interprets the halting of originally negative developments already as a success. But it was nearly impossible to achieve more far-reaching changes in the state of this problem in the short-term. Ozone-depleting substances remain in the atmosphere for several decades before finally loosing their destructive potential. Surprisingly, in this case the use of a standard which focuses on measuring relative improvements in the state of environmental problems comes to a less optimistic assessment than it would probably emerge from the use of the collective optimum as a possible baseline against which actual developments must be assessed. Analysts agree that significant steps were taken in this regime for reducing the production and consumption of ozone-depleting substances, even though one could object that these steps should have been taken much earlier.

A *worsening* in the state of problems has been observed for the biodiversity regime and also for the climate change regime. Developments in the state of problems in both regimes are not surprising. Measures which must be taken so that environmental problems will improve are complex and will only in the long run lead to desired improvements.

The question remains on whether observed levels of goal-attainment correlate with levels of problem-solving. In fact, such correlation exists particularly as far as environmental or economic goals are concerned. If major environmental or economic goals were not attained, no significant improvements in the state of related problems could normally be observed in regime elements. For example, failure to achieve goals which call for the sustainable use of components of biological diversity, for the conservation of biological diversity, or for regulation of safe handling and use of living modified organisms coincided with a negative trend in the state of the problem managed by the biodiversity regime. The non-attainment of a goal which aims to promote the production of tropical timber and timber products from sustainable managed sources finds expression in a further worsening of the environmental problem of increased tropical deforestation and rainforest degradation managed by the tropical timber trade regime. Finally, non-attainment of the climate change regime's ultimate objective to stabilize greenhouse gas concentrations in the atmosphere at a level that would prevent dangerous anthropogenic interference with the climate system coincides with a further worsening of the problem.

Did regimes have a causal influence on the state of the world? A scale from one to five has been used for measuring the causal impact of regimes on various dimensions of observed changes in the state of the world. Regime-factors and non-factors were distinguished in assessing the regime's causal impact. In making causal judgments about a regime's impact, it was also considered whether developments in the issue-area would have been different in the absence of a regime. The scale consists of a ranking that distinguishes between five possible forms of a causal impact of a regime. It defines a very strong causal influence and little or no causal impact of a regime as upper and lower limits on that scale. A very strong causal influence involves that regime factors account for virtually all developments. In addition, developments would have been very different in the absence of the regime. Little or no causal impact has been defined as a situation where non-regime factors mainly account for the state of the world or the situation would have not been different in the absence of the regime. A balanced causal influence occurred when regime and non-regime factors accounted equally for the state of the world. The absence of a regime would have led to a state of the world that partly differs from the observed situation. The two other ranks on this scale which define a significant or modest causal influence lie above or below the level of a balanced causal impact.

Correlations between data which describe the state of a problem and data which indicate the causal impact of a regime for problem-solving are made in Table 8.5. In 99 of the 187 problems for which data are available, the state of the problem improved slightly or considerably. The regime had a significant or very strong causal impact for these improvements in 47 of these 99 instances. This pattern pertains to problem-solving in 11 regimes. It was particularly distinctive in various elements of the Antarctic regime, the London Convention regime, the Rhine regime and the Barents Sea fisheries regime. In other regimes, the strong causal impact of a regime on problem-solving affected only a minority of regime elements. Regimes were not held responsible for a worsening in the state of the environment or for the observed failure to achieve improvements. For example, the climate change regime did not have a causal impact on the worsening of the problem of increasing greenhouse gas emissions.

Table 8.5 Causal Impact of Regimes on Changes in the State of Problems

Problem-Solving	Total of Problems		Little or no causal impact		Modest causal influence		Balanced causal influence		Significant causal influence		Very strong causal influence	
							Causal Impact of Regimes					
Problem worsened considerably	18	(9.6%)	8	(16.3%)	7	(17.1%)	0	(0.0%)	2	(4.3%)	1	(4.5%)
Problem worsened slightly	37	(19.8%)	22	(44.9%)	11	(26.8%)	2	(7.1%)	2	(4.3%)	0	(0.0%)
Problem stayed the same	33	(17.6%)	4	(8.2%)	10	(24.4%)	2	(7.1%)	13	(27.7%)	4	(18.2%)
Problem improved slightly	51	(27.3%)	8	(16.3%)	8	(19.5%)	13	(46.4%)	13	(27.7%)	9	(40.9%)
Problem improved considerably	48	(25.7%)	7	(14.3%)	5	(12.2%)	11	(39.3%)	17	(36.2%)	8	(36.4%)
Total	187	100.0%	49	100.0%	41	100.0%	28	99.8%	47	100.2%	22	100.0%

Summary

Our stock-taking which used two different approaches for exploring environmental problem-solving indicates that significant changes occurred in many environmental issue-areas. Empirical findings revealed that the broad majority of important goals which were established by regime members in an institution could be accomplished in the long-term. When states failed to achieve single goals in regimes, particularly far-reaching environmental goals were normally affected. More than one half of the problems coded for regime elements which end in 1998 experienced slight or considerable improvements. In some regimes, developments were less encouraging or even negative. About a quarter of our total of regimes was only determined by continued existence of the status quo, and for two regimes a negative trend continued during the 1990s.

Even though a few regimes were determined by developments that give cause for concern, the results which are available for the majority of regimes lead us to conclude that positive developments predominated. Causal judgments made by coding experts further suggest that most of the positive developments would not have occurred in the absence of a regime and that the causal impact of regimes was at least as relevant than external factors. Changes in the state of the environment are normally caused by a combination of different effects which are produced by regimes or by socio-economic factors that lie beyond the domain of a regime. The empirical results that emerged from the study of goal-attainment provided a more favorable picture about developments in the issue-area than the results that emerged from analyzing the evolution of the state of problems. Many of the goals that are established in global governance systems aim to improve the cooperative framework in environmental issue-areas. They are less difficult to achieve than those comprehensive goals which demand environmental problem-solving. For a number of regimes where watersheds distinguished several time periods in a regime, it became obvious that improvements in the state of a problem primarily occurred during those periods which ended in 1998.

Regimes, Programmatic Activities and Environmental Problem-Solving

Institutions are legal arrangements which aim to change a variety of behavioral patterns on the level of individuals, social groups, or international society. Regimes must be given institutional capacities which can contribute to overcome those social constellations that prevent effective problem-solving. Before changes in the state of environmental problems can be achieved, the programmatic activities carried out in regimes will have to produce a number of other changes in the social world. In the following, the three groups of regimes where slight or considerable improvements, continuation of the status quo, or a worsening of the state of a problem could be observed will be considered. The groups will be analyzed in terms of impacts of programmatic activities on the reduction of uncertainties in the cognitive setting or on compliance and whether observed changes had an impact on the state of environmental problems.

Cognitive Setting and Problem-Solving

It is a common assumption shared by institutionalists that institutional impacts on the cognitive setting or on compliance are necessary for achieving improvements in the state of an environmental problem (Ostrom 1990, 180; Young 1999b, 274–6). This assumption draws attention to the fact that more than one stimulus must normally come from an institution before changes in the state of the environment will occur. A correlation between data on the cognitive setting with data which illustrate changes in the state of environmental problems leads to the conclusion that progress in the knowledge base and the learning on the level of important states precede improvements in the state of the environment. Mostly, knowledge alone could not cause a re-definition of state preferences, but social action – whether taken by states or non-state actors – was required to transport this knowledge into the broader public, or to interpret and politicize it.

In 10 of the 16 regimes where slight or considerable improvements determined developments on the overall level of a regime by the end of the twentieth century, very strongly or strongly established understandings about the causes and effects of these problems could be found. In the other regimes of this group, this understanding had been a bit less advanced. In nearly each of these 16 regimes, information about policy options had reached at least a middle level of completeness. In some of the regimes where temporal watersheds distinguished different time periods of a regime element, it became obvious that in the early period of a regime the understanding about the nature of the problem was less advanced than in the period which ended in 1998. These cases demonstrate that a strengthening of international policies is usually closely linked to progress in the completeness of consensual knowledge. The work carried out in institutional mechanisms of regimes has mostly been intensified and broadened over time so that a growing number of causes and effects of a problem has been monitored and the completeness of consensual knowledge could be further increased. A case in point is the CITES-regime where problems like the maintenance of a sustainable and legal trade in plants and animals or the protection of endangered species were inadequately understood until the end of the 1980s. Strongly established understandings about the nature of these problems emerged in this regime only during the period between 1989 and 1998. The state of the two problems improved slightly during the earlier period, but improvements became more pronounced during the 1990s.[4]

---

4   It should be noted as a caveat that the actual purpose of the CITES-regime is limited to only one of a variety of possible factors which cause the extinction of wildlife. In assessing the impact of this regime, it has been argued by Peter H. Sand (1997, 26) that the regime is "not a general wildlife management treaty (whether it ought to be is

This development coincided with a further deepening of the density and specificity of regime rules. From the group of regimes where improvements in the state of the problems could be observed, developments in several regimes indicate that the existence of an advanced level of consensual knowledge led to a deepening or broadening of regime rules over time. Intensive efforts of monitoring and scientific research have been spent in the ozone regime. They were accompanied by reviews about the adequacy of existing policies, by reviews of national implementation and by efforts for enhancing capacities for implementation in countries with economies in transition and in the developing world.

Progress regarding the quality and comprehensiveness of knowledge in the issue-area did not always lead to improvements in the state of problems. Strongly established understandings about the nature of the problem or medium completeness of information about policy options could also be found in regimes where the state of problems remained the same or worsened. In regimes where the state of problems was determined by continuation of the status quo or by further worsening, improvements in the quality of issue-specific knowledge frequently emerged in the course of time. This may give cause for cautious optimism. At least some of the cases of the group where problems remained the same and no improvement could be observed qualify for possible improvements in the long-term. But advanced consensual knowledge did not in each of the regimes where the overall situation was determined by improvements lead to an expansion of regime rules so that that causes and effects would have been sufficiently addressed. Even in some of the issue-areas where improvements could be observed on the overall level of a regime, possible causes of a problem were not yet dealt with comprehensively enough or regime rules were not as deep as it would be required for achieving more far-reaching improvements.

This finding partly relativizes, though not refutes, the assumption of knowledge as a factor which can lead to improvements in the state of the environment. On the one hand, economic and environmental problems can compete in a regime to an extent which makes it impossible for the regime to do justice to both of these interests. This became obvious in the whaling regime where strongly established understandings emerged for the two problems of the conservation of whale stocks and the orderly development of the whaling industry, but where only the environmental problem could be resolved. The advancement of consensual knowledge is not sufficient for achieving improvements in the state of the environmental problem. On the other hand a number of other social problems that determine the constellation of interests among actors must often be resolved in regimes. These findings suggest that factors which are partly located outside the domain of a regime – e.g., the inertia of a problem, domestic factors, support for

---

another matter). As it stands, it is but one component of the existing patchwork of global and regional wildlife regimes, narrowly focused on the transnational trade issue, which is only one of the multiple threats to wildlife; and hence should be judged by its contribution to mitigating that particular threat".

regime policies by transnational coalitions – can influence the degree to which the state of environmental problems can be changed (Young 1999b, 271–4). If consensual knowledge had been less advanced, improvements in the state of a problem could normally not be observed.

Compliance and Problem-Solving

At first sight, no fundamental differences exist regarding observed levels of compliance between the three groups of regimes if only those regime elements are considered where the exploration period ended in 1998. Variance in the level of compliance can normally not sufficiently explain why different levels of problem-solving occur. Levels of compliance behavior vary within the different groups of regimes nearly to the same extent than between them. For the most part, behavior in these different groups of regimes was determined by compliance or by less severe forms of non-compliance. This finding suggests that compliance must be reached so that improvements in the state of a problem can be achieved. But less severe forms of non-compliance must not immediately threaten the problem-solving. Developments in two regimes where watersheds distinguish two different time periods illustrate that improved compliance behavior in the period which ends in 1998 coincides with improvements in the state of the environmental problem. This finding applies to the whaling regime and to the intentional oil pollution regime. Occasional non-compliance by single states determined the early period of the whaling regime which has seen a considerable worsening of both problems managed by this regime. Behavior that meets regime requirements characterized the second regime period between 1982 and 1998. A considerable improvement occurred concerning the conservation of whale stocks, but no satisfactory solution was found for the regime's economic problem of an orderly development of the whaling industry. Regulations under OILPOL were insufficiently complied with between 1954 and 1978. Consequently, the problem of marine oil pollution from intentional discharges and tanker accidents worsened considerably during this period. Remarkable improvements of the problem were observed during the regime's second period under the components of MARPOL and regional MOUs. Strengthened efforts had been made "to improve government reporting on reception facilities and enforcement, to increase government enforcement of treaty regulations against tanker operators and tanker owners, and to get governments to ensure that ports have reception facilities for oil waste generated by ships" (Mitchell 1994, 121).

The existence of full compliance alone can not sufficiently guarantee that goals can be fulfilled and problems be solved. For example, insufficient compliance by single states with CCAMLR-regulations in the Antarctic Treaty System improved to a behavior that meets regime requirements during the period from 1989/91 onward. A less satisfactory level of compliance had prevailed during the 1980s.

During the latter period of the regime, CCAMLR's conservation problem did not change at first but stayed the same until the end of the exploration period.[5]

Our compliance data reveal that non-compliance is normally a problem that becomes relevant with respect to single states rather than a behavioral pattern that determines the overall situation in a regime. Less severe forms of non-compliance by single states could occasionally be observed also in the group of regimes were problems improved slightly or considerably. This indicates that compliance mechanisms are normally inevitable, because temporary problems of single states to comply with regime rules could otherwise become permanent and erode the functioning of the institution as a whole. Regime members did not tolerate non-compliance in the long term. Existing institutional mechanisms provided an arena where compliance problems could be discussed and possible solutions be developed. Compliance mechanisms could contribute to improving compliance by single states partly in combination with measures for capacity-building.

State Preferences and Problem-Solving

By promoting the evolution of consensual knowledge, regimes contributed to the fact that single states no longer denied the existence of a problem or avoided to raise serious doubts concerning possible causes and effects. Some examples can illustrate that state preferences have changed towards more regime-conducive behavior. These changes were crucial so that the worsening of a problem could be stopped or that the state of the environment improved in the long term. These examples can be obtained from those regimes where watersheds distinguished different time periods. They allow comparison of the attitude of an important state towards the character of regime policies between different time periods. During the second half of the 1980s, a change occurred in the attitudes of some European states which had initially been opposed to the creation of international policies for the reduction of ozone-depleting substances. The change which could be observed in France or the United Kingdom was supported by the findings of combined monitoring and research initiatives made under the leadership of research institutions in the U.S. and of international organisations which improved the knowledge about the causes and effects of ozone depletion in the stratosphere (Breitmeier 1996; Litfin 1994; Parson 2003). Later on in the 1990s, these countries finally complied with (or sometimes over-complied) regulations of the ozone regime together with Japan and other industrialized countries which had originally been against reduction measures. A similar example is the change of Germany's attitude towards a reduction of air pollutants in Europe which could

---

5   Some developments that occurred in this regime component during the 1990s suggest that "the emergence of cooperative multi-species stock surveys, precautionary regulations based on ecosystemic considerations, and the formal structure at least of a credible inspection system" gave prospect for improving ecosystemic conservation in this component (Stokke 1996, 151).

be observed in the early 1980s. Germany's change in position was not only caused by the evolution of scientific consensus Other factors like rising public awareness or the emergence of environmental groups also facilitated change in Germany's viewpoint on the issue (Cavender-Bares, Jäger and Ell 2001).

Such far-reaching positive impacts of consensual knowledge on state attitudes could not always be observed. In a few cases, laggard states which were opposed to the development of effective policies during an earlier period stayed with their negative attitudes also in subsequent regime periods although the completeness of consensual knowledge had improved. For example, Greece took a negative position regarding the development of more effective policies in the oil pollution regime during the early and later periods. But the regime has been ascribed a large causal impact for an improved (though not full) level of compliance by this country during the second period. In the whaling regime, Norway and Japan are rare examples which demonstrate that a state's attitude towards regime policies can become more negative during a later period of political management. Both countries had rather negative attitudes vis-à-vis the basic character of the regime during the second period. But the regime has been ascribed a large causal role that these countries had complied for the most part with its rules also during this period. These critical cases demonstrate that even though state attitudes must not inevitably correspond with the goals of a regime, institutional mechanisms can prevent that single states will weaken their efforts to comply with regime rules. The question remains on whether observed compliance behavior always reflects hidden motivations held by states in an issue-area. The prominent case of Norwegian and Japanese behavior in the whaling regime may remind us that actual behavior must not always coincide with national preferences. Although both countries were not successful in their efforts of loosening the ban for commercial whaling, they remained in the regime. Obviously, the dynamics in the whaling regimes as well as external factors such as political and economic pressure or the activities of the NGO-community also accounted for this behavior (Andresen 2002, 396–9).

While institutions have been important for the production of regime-conducive state attitudes in single cases, our data on the role of important nations indicate that, for the most part states which were sympathetic to the evolution of international policies during earlier periods also maintained their regime-conducive attitudes during later periods of a regime. This is understandable when states are severely affected by an environmental problem or have a special interest in the sustainable management of shared natural resources. But this finding also illustrates that active participation by a state in a regime can result in the long-term stabilization of its preferences or political behavior in the issue-area. The role of a pusher state which supports the development of far-reaching policies frequently involves participation in the management of various programmatic activities. Long-term participation of a state in international programs for scientific monitoring, research about cause-effect relationships, or technological assistance leads to the creation or deepening of transnational networks among scientists and policy experts and strengthens the bonds between national institutions and a regime. This can partly explain why many states maintained their positive attitudes within regimes in the long term despite observed changes of governments or the evolution of competing interests on the domestic level. When pusher states had entered into a special commitment in a regime which involved to support the fulfillment of its goals and to actively participate in the management of regime functions, they normally accepted the path-dependency combined with this commitment.

A significant number of important states remained neutral or skeptical concerning the development of effective policies throughout the different time periods of a regime. In some instances, states had taken this neutral position during the latter period of the regime after their original position had been even more negative. Neutral positions had been taken by single states during the management of regimes even when the knowledge about cause-effect relationships or information about policy options had proliferated. In some regimes, developing countries or countries with economies in transition accepted the findings of scientific assessments regarding the causes and effects of a problem. However, they also indicated that they could only contribute to international problem-solving if they would be given those capacities which were required for implementation, or if international policies would be less far-reaching than intended by pusher states. The tension between arguments which refer to consensual knowledge or which are based on economic interests has occasionally also determined the attitudes of industrialized countries. In some cases where they feared negative impacts of regime policies on domestic interest-groups, industrialized countries adopted an attitude of wait and see or continued their opposition during negotiations against the implementation of more far-reaching environmental policies. This illustrates that the willingness by states to obey international policies does frequently not purely emerge from arguing and persuasion, but that the readiness of states to accede to and implement environmental treaties also depends on the ability of an institution to avoid that implementation will cause severe financial or other burdens for a state.

Non-State Actors and Problem-Solving

Our approach on the role of non-state actors helps to further understand the ways through which institutional mechanisms cause change in the social world. Non-state actors are perceived as contributors to the achievement of functional and environmental goals during regime management (Oberthür, Buck and Müller 2002). They are also understood as advocates of environmental, economic, or social interests and as supporters of ideas in world society which pressurize the international community of states (Keck and Sikkink 1998). This pressure can consist of support for, or opposition to, environmental policies.

Before embarking on further analysis, it should be noted that the database lacks information about the depth of involvement by these actors in a regime's programmatic activities. Likewise, the role of political pressure by interest-groups can not purely be illustrated with reference to quantitative data. In addition to quantitative findings about the participation of non-state actors in regime management or about their role during agenda-setting or negotiations, reference will also be made to narratives written by coding experts on the topic. Further on, a few qualitative case studies will be used to illustrate the role of non-state actors in particular regimes.

Non-State Actors and Regime Management

Our distinction between functional and environmental goals took into account that goals are causally connected with one another. The fulfillment of functional goals is normally a precondition for achieving environmental goals or for changes in the state of environmental problems. Thus, an explanation about changes in the state of environmental problems is normally composed of a more comprehensive causal chain than an explanation which concentrates on the fulfillment or non-attainment of functional goals. These goals require the implementation of more or less clearly defined tasks by an institution. The functional goals that existed in regimes normally found expression in the creation of institutional mechanisms. Occasionally, non-state actors participated on the domestic level in managing these tasks already before an international regime emerged. Public-private partnerships on the domestic or international level in some way contribute that states are able to fulfill their tasks especially in times of a rapidly changing socio-economic environment (Huckel, Rieth and Zimmer 2007). The state responded to the functional differentiation of policymaking, to the decreasing ability of the state for autonomous problem-solving, or to increasing complexity of political issues within and beyond national boundaries by strengthening the creation of networks with domestic and transnational actors for the purpose of safeguarding its political steering capacity (Teubner 1999). International governance systems faced a similar demand for growing inclusion of networks of non-state actors, but it is doubtful whether inclusion of these actors has already constrained the scope for action of governments. Governments which can build on domestic networks can raise their influence in regimes by linking the activities of domestic experts closely to the work of programmatic activities. Regime bodies that were established for the management of functional tasks or regime secretariats kept these activities under control so that networks of experts would not become too independent from the will of states.

It was demonstrated in the chapter dealing with changes in the knowledge-base (see Chapter 6) that networks of scientific, legal or policy experts actively participated in political processes in a broad majority of regime elements. For the most part, these contributions by non-state actors affected the implementation of goals which demanded to improve scientific collaboration and research in the issue-area, to monitor causes and effects of a problem or state behavior, to improve implementation on the domestic level, or to further develop technical and economic options that can be used for making the implementation of international rules more effective. But qualitative studies also illustrate that efforts to include non-state actors in domestic implementation have not always led to desired results. Summarizing the findings of a number of comparative case studies which deal with the implementation of international environmental regimes, Kal Raustiala and David G. Victor (1998, 665) concluded that "the general proposition that participation matters is not sustained in all cases" and that participation "has clearly had an influence on declared policies, but not always on the behavior that follows". In particular, the assumption that participation of target groups could motivate them to implement legal agreements more thoroughly could not be verified by each of the cases explored in connection with the above-mentioned comparative case studies. However, participation by target groups clearly led to the provision of implementation expertise consisting of better information "on the range of possible policy options, technical feasibility, and costs and benefits" (Raustiala and Victor 1998, 666).

It is often difficult to separate participation in regime management from participation in domestic implementation. The study of goal-attainment that primarily refers to the level of a regime suggests that many goals in regimes could not have been attained without the support of non-state actors. For example, several important goals established in the Barents Sea fisheries regime could only be achieved by including non-state or semi-governmental actors. As a result of nongovernmental research cooperation between the Norwegian Institute of Marine Research (IMR) and the Russian Polar Research Institute of Fishery and Oceanography (PINRO) the coordination of marine research and of research related to fisheries issues in the Barents Sea has improved (Stokke, Anderson and Mitrovitskaya 1999, 115–21). The inclusion of fisheries organizations or private firms from both countries in the regime made it possible that policies of the regime were obeyed by private actors. Various important goals of the Great Lakes management regime required that non-state actors from the U.S. and Canada actively participated in the programmatic activities of the regime. These long-term goals inter alia involved developing and adopting a systematic and comprehensive ecosystem approach to the management of the Great Lakes or to develop and implement Remedial Action Plans (RAPs) and lake-wide management plans to restore beneficial uses of Great Lakes waters. The inclusive institutional structure of the Great Lakes management regime made it possible for local decision-makers, national agencies, industry, research institutes and other non-state actors to become engaged in a process where problems in the area of concern were commonly identified, where public education about the causes and consequences of environmental problems could be strengthened, technological tools and techniques for pollution-prevention be identified, or where scientific research and assessment or monitoring could be strengthened. In assessing the achievements of the Great Lakes Water Quality Agreement, a team of authors concluded that a "strong and diverse nongovernmental community developed as a result of the formal structures, and the formal structures have been energized and legitimized because of the continuing active involvement of that community"

(Valiante, Muldoon and Botts 1997, 217). The attainment of goals which aimed to improve the understanding about the nature of the problem or the development of policy options benefited from the work of the International Association of Great Lakes Research (IAGLR) and from active participation of other non-state actors such as the Sierra Club or Great Lakes United (GLU) as a broad bi-national coalition of nongovernmental organizations. Participation by industry and bi-national industrial organizations such as the Council of Great Lakes Industries has been relevant for the development of feasible policies.

Many other examples could be referred to which can illustrate that non-state actors contributed to the achievement of functional goals in regimes in various ways. Such examples have been provided in previous chapters where it could be demonstrated that these actors contributed to compliance management (see Chapter 7). The CITES-regime for example has been referred to as case in point where compliance monitoring has strongly benefited from the activities of TRAFFIC as a joint monitoring network established by WWF and IUCN and which works in collaboration with the CITES-secretariat. International organizations provided their scientific and technical services particularly in multilateral regimes. In some of these regimes, they were partially successful with their efforts to expand the involvement in technical and scientific activities that reached beyond their original domain.

It would be exaggerated to argue that failure to achieve functional goals or lacking improvements in the state of the environment are primarily caused by insufficient inclusion of non-state expertise. Efforts have been made in regimes to strengthen the input of non-state expertise, but the contribution of these actors is not a universal remedy that will guarantee the attainment of functional or environmental goals in any case. The climate change regime can be used as an example which illustrates that many activities of non-state actors occur long before changes in the state of a problem can finally be achieved. During the initial phase of the regime, member states had to pay special attention to advance the consensual knowledge and to expand and improve the functioning of institutional mechanisms before concrete policies on emissions could be negotiated. Kal Raustiala (2001) illustrates that the IPCC and other scientific actors, policy experts or economists played an important role in the climate change regime. Nevertheless, the author concludes that "much of this NGO activity largely comes not at the expense of state power, but rather to the mutual advantage of states and NGOs" and that the "participation of NGOs in formal international cooperation such as the FCCC enhances the ability, both in technocratic and political terms, of states to regulate new areas through international agreements" (Raustiala 2001, 115). Non-state actors can hardly serve as a starting point for explaining why reduction measures in industrialized countries which could lead to a slowdown of greenhouse gas emissions will only take effect more than a decade after the negotiations about the Kyoto-Protocol have been brought to an end. This illustrates that in addition to non-state actors other factors are finally responsible for achieving improvements in the state of the environment. Non-state actors alone could not change the state of the world for the better – but without their expertise many functional goals would probably not have been attained.

Some of the functional goals can only be attained in the long term or involve the management of programmatic activities on a regular basis so that they require enduring participation by single non-state actors. Long-term involvement of non-state actors in the management of regime functions can be observed in the field of scientific research or monitoring where the Scientific Committee on Antarctic Research (SCAR), the IPCC, IIASA, IUCN, Wetlands International or Birdlife International became involved in regimes on the Antarctic, climate change, long-range trans-boundary air pollution, or the conservation of wetlands. Nevertheless, some regime functions can involve that participation of single non-state actors in policy-networks will be temporary. For example, possible causers of a problem can be invited to participate in transnational policy-networks which deal with the development of practical solutions in the issue-area. This participation can end when measures have been implemented and actors can no longer contribute to the fulfillment of a regime's functional or environmental goals.

Social Change, Goal-Attainment and Problem-Solving

Many students of world politics assume that non-state actors are driving-forces for the production of social change. Political pressure is frequently used as a variable that contributes to broaden the functional scope of regime policies or their density and specificity. How far were processes like agenda-setting or negotiations determined by non-state actors? Which examples exist in our data-set that can illustrate the role of non-state actors for raising transnational public awareness about environmental problems or for changing the attitudes and behavior of states? Our first measurement determined whether the inclusion of issues on the political agenda of regimes has emerged from the influence of a single state or a small group of states, from a deliberate process of negotiations among more or less equal states setting the terms of the agenda deliberately, or from factors largely outside deliberate efforts of potential regime members (e.g. external shocks, nongovernmental actors). Independent measurements have been made for each of these three factors on a scale from one to five with regard to their relevance for agenda-setting in a regime element. The upper limit on that scale referred to a situation where agenda-setting had been strongly determined by this factor, whereas the lower limit on that scale pertained to a situation where this factor has been irrelevant. The middle of that scale covered a situation where this factor could be ascribed some relevance but where agenda-setting was only partially determined by these factors. Ranks located directly above and below the middle pertained to situations where these factors were less pronounced than the upper limit or more relevant than described by the lower limit of the scale.

First of all, the findings that are included in Tables 8.6, 8.7 and 8.8 reveal that states had a strong influence on the composition of political agendas in regimes. In 60 (43.5 per cent) of a total of 138 regime elements the inclusion of issues on the

Table 8.6   Influence of a State or a Small Group of States on Agenda-Setting in Regime Elements

Was the Inclusion of Issues on the Agenda Determined by a Single State or a Small Group of Potential Regime Members?	All Regime Elements		Regime Elements which end in 1998	
1 = Very strongly dominated by a single state or a small group of potential regime members	17	(12.3%)	8	(9.6%)
2 = Strongly dominated by a single state or a small group of potential regime members	43	(31.2%)	25	(30.1%)
3 = Relevance of domination by a single state or a small group of potential regime members	31	(22.5%)	18	(21.7%)
4 = Minor relevance of domination by a single state or small group of potential regime members	35	(25.4%)	24	(28.9%)
5 = Domination by a single state or a small group of potential regime members not at all relevant	12	(8.7%)	8	(9.6%)
Total	**138**	**(100.1%)**	**83**	**(99.9%)**

Table 8.7   Agenda-Setting as an Interstate Process

Was the Inclusion of Issues on the Agenda Determined by Potential Regime Members Negotiating Among Themselves More or Less as Equals and Setting the Terms of the Agenda Deliberately?	All Regime Elements		Regime Elements which end in 1998	
1 = Very strongly an interstate process of potential regime members negotiating among themselves	38	(27.5%)	23	(28.0%)
2 = Strongly an interstate process of potential regime members negotiating among themselves	58	(42.0%)	38	(46.3%)
3 = Relevance of an interstate process of potential regime members negotiating among themselves	35	(25.4%)	20	(24.4%)
4 = Minor relevance of an interstate process of potential regime members negotiating among themselves	7	(5.1%)	1	(1.2%)
5 = Interstate process not at all relevant	0	(0.0%)	0	(0.0%)
Total	**138**	**(100.0%)**	**82**	**(99.9%)**

agenda was strongly or very strongly determined by a single state or a small group of states. In 96 (59.5 per cent) of 138 regime elements the inclusion of issues on the agenda was strongly or very strongly determined by an interstate-process where potential regime members were negotiating more or less as equals and setting the terms of the agenda deliberately. Secondly, external factors like sudden shocks or

Table 8.8   The Influence of External Factors on Agenda-Setting

Was the Inclusion of Issues on the Agenda Determined by Factors Largely Outside Deliberate Efforts of Potential Regime Members?	All Regime Elements		Regime Elements which end in 1998	
1 = Very strongly determined by factors outside deliberate efforts of potential regime members	8	(6.1%)	7	(9.0%)
2 = Strongly determined by factors outside deliberate efforts of potential regime members	40	(30.3%)	20	(25.6%)
3 = Relevance of determination by factors outside deliberate efforts of potential regime members	38	(28.8%)	29	(37.2%)
4 = Minor relevance of determination by factors outside deliberate efforts of potential regime members	27	(20.5%)	14	(17.9%)
5 = Determination by factors outside deliberate efforts of potential regime members not at all relevant	19	(14.4%)	8	(10.3%)
Total	**132**	**(100.1%)**	**78**	**(100.0%)**

non-state actors could influence the inclusion of issues on political agendas. In 48 (36.4 per cent) of the total of 132 regime elements these factors determined the agenda-setting strongly or very strongly. The impact of external factors on agenda-setting amounted in a number of regime elements to the influence of states. But in none of the cases they were more important than the state influence. The most important impact on agenda-setting seems to have emerged from inter-state activities. The data that describe developments for regime elements which end in 1998 indicate that inter-state activities were ascribed a very strong or strong relevance for inclusion of political issues on the agendas of regimes in 61 (74.3 per cent) of a total of 82 regime elements. This finding suggests that political agendas in environmental regimes have still been more determined by states than by non-state actors, even though one has to admit that these actors were relevant in most of the regimes explored.

Even when agenda-setting has been strongly determined by external factors, states were always ascribed similar relevance. These findings also lead to the conclusion that non-state actors often focused their activities solely on specific regime elements. In regimes with various institutional components non-state actors focused their activities in some cases primarily on those issues which they considered particularly relevant, whereas they avoided giving each regime issue the same priority. A case in point is the engagement of environmental non-state actors against the implementation of the Convention for the Regulation of Antarctic Mineral Resource Activities (CRAMRA) during the late 1980s and early 1990s. The data provided by both coding experts suggest that state influence has been more influential than external factors for agenda-setting in most of the components of the Antarctic regime. But in combination with the activities of Consultative Parties like Australia or France which raised doubts as to whether CRAMRA would serve their own purposes and consider environmental concerns non-state actors and growing environmental awareness were considered responsible that transnational discourse increasingly focused on CRAMRA's negative impacts on the environment. As a result, regime members renounced exploitation of Antarctic minerals and added the Environmental Protocol to the Antarctic Treaty System in 1991.[6]

Taking a closer look on the relevance that was ascribed to single types of actors for agenda-setting confirms the above-described findings. In exploring the presence and relevance of various types of actors on the agendas of regimes during regime periods which end in 1998, the finding emerged that the leadership of state pushers or of a hegemon in the issue-area played a role in 86 of a possible total of 102 regime elements – whether these factors have only been present or where considered most influential for agenda-setting. In 85 regime elements, various types of non-state actors such as activist nonprofit interest groups, industrial organizations and multinational corporations, scientific organizations or epistemic communities were identified as present or most influential for agenda-setting. In 53 regime elements, at least one of these different non-state actors has been described as most influential for agenda-setting – mostly in combination with other important factors like state influence. More than a single type of non-state actors has been frequently identified as present and active during agenda-setting. This finding gives reason to assume that nonprofit activist interest groups are not the only important type of actor for the production of social change, but that various types of non-state actors participated together with states in the framing and politicization of transnational issues.

This finding draws attention to the fact that mobilization for support or opposition to the formation or further evolution of regimes frequently emerges from collaboration between important states and non-state actors. For example, a combined view on those data which describe the roles of states or non-state actors as pushers or laggards in a regime reveals that the United States could count on the support of other states and of activist or service-oriented non-state actors in each of the 10 regimes where this country predominantly took a role as a pusher for the strengthening of international environmental policies. IUCN, WWF or Friends of the Earth have played such a supportive role in more than one of these regimes.

---

6   Christopher Joyner (1996, 165) illustrates that CRAMRA's demise came precipitously and was surprising "considering the extraordinary investment of time, energy and intellectual effort made by ATCP governments in the minerals negotiations during the 1980s". The answer to the question about why CRAMRA had been dumped by ATCPs although it had been a painstakingly negotiated and highly sophisticated legal instrument lies "in perceptions of CRAMRA's ineffectiveness, together with a new-found conviction by the Consultative Parties that the mineral treaty's purpose was neither appropriate nor desirable for the Antarctic".

Together with other states and non-state actors the United States and Greenpeace were both supporting policies in the two regimes dealing with the conservation and management of tuna and tuna-like species in the Atlantic and the Eastern Pacific, in the whaling regime, in the oil pollution regime, or in the ozone regime. Even though both actors did not always fully agree on the character of policies which they regarded as necessary in an issue-area, the fact that a powerful state and an ideologically powerful transnational actor both followed similar political goals involved that these topics could remain on the international agenda. At same time Greenpeace and the United States (or other industrialized countries) were acting as strong political opponents during the formation and further evolution of the 1989 Basel Convention on the Control of Trans-boundary Movements of Hazardous Wastes and their Disposal (Krueger 1996, 8–13). The laggard position of the United States was backed during the evolution of the hazardous waste regime by economic interest-groups such as the United States Chamber of Commerce or the Bureau for International Recycling. This illustrates that non-state actors are not considered as disturbing factors by states. They are used by states for the building of tacit coalitions which put forward similar ideas in world politics – whether these ideas support or are opposed to environmental policies.

Non-state actors had a more significant impact on the inclusion of issues on the political agenda than on negotiations. For 22 regimes, data are available which describe whether negotiations were determined by single states, by an inter-state process of bargaining, or by transnational forces. In the broad majority of these regimes, transnational forces played a less important role. This result is only partly caused by the fact that these data frequently pertain to negotiations about framework conventions or regime components which took place until the early 1970s before non-state actors became more active on the transnational level. Even in a number of those regimes where non-state actors played important roles as pushers of environmental policies on the political agenda, they had less influence on negotiations than one would expect.

The disparity that frequently occurs between the impact of activist non-state actors on agenda-setting and their less influential role for negotiations can be illustrated by referring to a number of narratives made by individual coding experts about the formation and evolution of regime elements. With respect to the formation of the biodiversity regime, it has been argued that "NGOs and expert communities were centrally important in the onset of the issue and in raising it on the international agenda but less influential later".[7] Other narratives on regime formation made by experts indicate that non-state actors were contributing to place the issue on the international agenda or contributed to re-frame issues in more environmental terms. For example, both coding experts agreed that during the early period of the regime on wetlands a handful of NGOs was important "in providing leadership, momentum, and work to the process".[8] In the late 1950s,

---

7   Narrative on aspects of regime formation made by Kal Raustiala.
8   Narrative on aspects of regime formation made by Dwight Peck.

the International Waterfowl Research Bureau (IWRB), the International Union for the Conservation of Nature and Natural Resoruces (now IUCN – The World Conservation Union) and the International Council of Bird Protection (ICBP) started a joint project which was intended "to increase awareness of the importance of wetlands to humanity and contribute to their conservation".[9] Subsequent conferences that were partly organized by non-governmental organizations dealt with various proposals for a convention on wetlands and non-state actors actively participated in drafting these texts. Likewise, agreement by two coding experts on the productive role of non-state actors during the early period of a regime can be found in the narratives provided about the formation of the tropical timber trade regime. The initial intention to establish a commodity agreement on tropical timber had its origins in the international commodity negotiations sponsored by UNCTAD in 1976, but the environmental problem of accelerated deforestation in tropically forested countries was only considered in the early 1980s. Under these circumstances, the activities of non-state actors were considered important so that the ITTA evolved from a proposal for a commodity agreement to a hybrid commodity-and-environment agreement by the time it was finalized".[10] Further on, several non-state actors like the IIED, IUCN, WWF and Friends of the Earth which felt that the ITTA "would provide an important forum to discuss forest management practices in tropical timber countries" hosted a conference in 1983 to convince governments to ratify the agreement.[11] Some textual responses also indicate that the activities of states during agenda-setting or negotiations can be caused by domestic pressure of non-state actors or other factors which can be found on the domestic level. For example the component which has been developed for the conservation and management of dolphins in the IATTC-regime "was primarily initiated by the USA as a result of their domestic legislation, which was motivated by public concern expressed through NGO environmental groups".[12]

Conclusion

The empirical findings provided in this chapter suggest that the majority of goals established in regimes were accomplished in the course of time. In some regimes, the non-attainment of functional or environmental goals could be observed. The state of environmental problems improved in the majority of regimes, but there are still regimes where developments were less encouraging or even negative. Nevertheless, the coding revealed that institutional mechanisms were important for achieving improvements or for avoiding a further worsening in the state of the environment. In some respects, environmental problem-solving can be understood

---

9   Narrative on aspects of regime formation made by Clare Shine.
10   Narrative on aspects of regime formation made by Debbie Davenport.
11   Narrative on aspects of regime formation made by Fred Gale.
12   Narrative on aspects of regime formation made by James Joseph.

as a process where several functional goals must be fulfilled before improvements in the state of the environment will be reached. Under the circumstances, environmental regimes established goals which provide that uncertainties in the knowledge base will be reduced, access to information about compliance behavior be improved, compliance management be strengthened, or exchange about measures chosen for implementation on the domestic level be enhanced. To this end, programmatic activities have been established in regimes to bring forward the fulfillment of these goals. One could argue that the study of goal-attainment draws too much attention to whether the internal logic of policy-makers will be fulfilled whereas it avoids to assess whether these goals meet the requirements for environmental problem-solving that exist against the background of a collective optimum. This argument is based on the assumption that the goals established by policy-makers are often less far-reaching than the goals put forward by external analysts or environmentalists. This may be true in single cases. However, the empirical study revealed that the framing of environmental goals in regimes is normally more far-reaching than skeptics may expect. In addition, it became obvious that the number of goals can increase within regimes or become more demanding over time when new institutional arrangements are concluded that consider the need for more comprehensive action.

The above-described findings suggest the conclusion that non-state actors support and complement the activities of states in international regimes. But they did not replace the role of the state in world politics. In many cases, it is still difficult to draw a simple correlation between the scope and depth of non-state actor involvement on the one hand and goal-attainment or the degree of problem-solving on the other hand. While functional goals could frequently only be attained by institutions with additional support of non-state actors, the attainment of environmental goals or environmental problem-solving are also dependent on various other factors. During agenda-setting and processes that occurred outside negotiations non-state actors were more influential than in negotiations where states maintained their authority to negotiate on issues. Nevertheless, single cases also illustrate that the work of programmatic activities should not be considered separately from political processes taking place in the transnational public sphere. Activist environmental non-state actors can help to detect and politicize noncompliance of states and thus make the work of programmatic activities more effective.

Obviously, the contribution of non-state actors to goal-attainment or problem-solving did not fundamentally constrain the traditional role of the state as an agent of national interests on the international level. Nevertheless, the expansion of functional goals and of regime functions caused a growing demand for the support of non-state expertise during the management of regimes. The production of social change and the attainment of goals or improvements in the state of the global environment emerge from the interaction of various types of non-state actors. Some of these actors actively participate in regimes where they contribute to improve the scientific knowledge, provide information about compliance behavior, or participate in the development of international policies and domestic implementation. Others operate predominantly outside the domain of a regime but influence societal preferences to that effect that governments will be forced to adopt more far-reaching measures. The question remains on whether goals would also have been achieved if non-state actors had been less active in the transnational public sphere and if states would not have been able to resort to the potential which many service-oriented non-state actors can offer for goal-attainment or problem-solving. The state lacks many of the resources which are necessary for managing problems effectively. Thus in conclusion, the findings suggest that regime management has considerably benefited from participation of non-state actors and that the support of these actors in the transnational public sphere has been a valuable resource which states could use for achieving their goals in international politics.

Chapter 9

Distributional Consequences of Environmental Regimes

The distribution of values that arise from an institution can impact the willingness of states or social actors to obey international rules. Conflict about the equitable character of distributional consequences has become widespread in international institutions.[1] Industrialized countries were confronted with demands to support the special needs of developing countries or countries with economies in transition to implement policies of environmental regimes, but the character of distributional consequences has also been raised as an issue in relations among industrialized countries. The debate over distributional consequences was not constrained to the level of the international state system. Global governance systems faced also the challenge to justify their distributional outcomes vis-à-vis social groups such like consumers, car drivers, wood cutters in tropical forests, workers, trade unions, industrial firms or environmental NGOs.

Before we embark on the empirical investigation, a caveat must be made. Normative criteria for determining the equitable character of social order have been established by political theorists primarily for the level of the nation-state. But on the international level, the understanding of distributional justice has been less clearly developed. The debate about the responsibilities for the increase of greenhouse gases in the atmosphere and about the negative effects of climate change illustrates that justice is an important issue not only on the domestic but also on the international level (Vanderheiden 2008). It should also be kept in mind that the analysis of distributional consequences can not be equated with the analysis of the equitable character of regime impacts. Before it can be assessed whether regime consequences correspond to normative criteria of equity, the consequences themselves have to be identified. This chapter will primarily focus on analyzing the distributional consequences of regimes. In the following, it will first be explored whether the issue of distributional justice has been considered in the legal provisions of regimes and which distributional consequences have been produced by regimes. Second, it will be explored whether regimes established institutional mechanisms through which distributional consequences could be

---

1 For example, the world trade regime has been used by developing countries for pressurizing industrialized countries to reduce existing economic asymmetries between North and South. During the 1970s and 1980s, developing countries made efforts in commodity trade negotiations to change the commodity regime that emerged after World War II (Finlayson and Zacher 1988).

influenced to achieve impacts that consider different levels of socio-economic development of states. Third, it will be explored which relevance can be ascribed to non-state actors for changes in the distributional impacts of regimes.

For a number of reasons, the intercoder-reliability of data will not be analyzed in this section. First, agreement about the existence of basic principles and norms in a regime had been reached between the two experts coding a regime before the actual coding had begun. Second, analysis of the impact of institutional mechanisms on distributional consequences is mostly based on those data describing the programmatic activities of a regime for which agreement had also been reached before experts had finally taken up their coding work. Third, a great portion of the information about distributional consequences has been provided in the form of written narratives. Most of the narratives made by two experts for a regime complement one another. These narratives were mostly provided for the regime as a whole rather than for single regime components. Therefore, no distinction will be made in this section between those cases which ended in 1998 or before.

Measurements do not reflect the accuracy of studies carried out in the field of resource economics. In some of these narratives experts reported their uncertainties regarding the knowledge that was available about the distribution of possible costs and benefits. For example, it has been indicated for the biodiversity regime that "it is very difficult to undertake an overall cost-benefit-analysis as there is no comprehensive valuation of biodiversity goods and services which include a social, economic and ecological assessment" and that "methodologies to valuate biodiversity goods and services are complex and contentious".[2] For the climate change regime it has been indicated that environmental benefits could, if at all, hardly be identified by the end of the twentieth century. It has been argued that "as the regime has yet to yield direct 'benefits' in terms of greenhouse gas emissions, it is really only through the operation of its financial mechanism that concrete costs and benefits can be assessed".[3]

The Evolution of Legal Norms and Distributional Consequences

How far has the issue of justice been considered as a relevant topic by states within regimes? Which distributional consequences did emerge from environmental regimes? Two different measurements will be made to answer these questions. First of all, environmental regimes will be examined with regard to whether they include explicit or implicit principles or norms which prescribe that distributional justice shall be considered in developing and implementing regime policies. This stocktaking can inform about whether international society acknowledged claims

---

2   Written narrative about the distribution of costs and benefits made by Gudrun Henne.

3   Written narrative about the distribution of costs and benefits made by Jacob Werksman.

for justice and sustainability as relevant ideas. Second, analysis will explore the costs and benefits that emerged for states in environmental issue areas as a result of the existence of a regime. It will be assessed whether distributional consequences reflect differentiated responsibilities for the causation of a problem, whether they are based on legitimate claims for the use of environmental goods, or whether they consider economic and financial capacities that are required for problem-solving.

Norms of Justice

International environmental law developed various legal principles which demand that international policies shall reflect equitable solutions (Beyerlin 2000). These norms primarily pertain to states as those subjects which constitute international society. Information about the existence of these principles and norms (whether they existed singly or in combination with other principles or norms of justice) can illustrate whether distributional justice has been acknowledged as an issue by regime members. Our data protocol included a nominal scale that was used for identifying the existence of basic legal principles and norms in regimes. The list considers three types of basic principles and norms that emerged in international environmental law (Sands 1995; Sands, Tarasofsky and Weiss 1997): i) *basic norms of international society* confirm the sovereignty of the nation-state or prescribe good neighborliness and cooperation among states; ii) the list considers *basic environmental principles* that emerged during the past few decades; iii) five international principles and norms were considered which cover various aspects of *equity and sustainability*.

In some respects, the five principles and norms of equity and sustainability overlap. The "principle of the common heritage of mankind" gained importance during the evolution of the Law of the Sea and of other international agreements which govern the use of global goods (Joyner 1999). The "principle of equitable use of natural resources" calls for equity between states that are at different economic levels, that have different environmental and developmental needs, or that contribute to particular problems to different extents. The "principle of sustainable use of natural resources" focuses on the adoption of standards which govern the rate of use or of exploitation of specific natural resources. Since it implies that catches can be limited to maximum sustainable levels for natural resources such as fish-stocks, it intends to protect present generations from the negative effects of egoist utility-maximizing behavior of actors as well as future generations from the depletion of these resources. The principle of "integration of environment and development" involves a commitment to integrate environmental aspects into economic development and to consider economic and social needs in determining and implementing rules for environmental protection. Finally, the "principle of intergenerational equity" emphasizes the responsibility of members of present generations for future generations.

For a total of 21 regimes data are available that describe the most important principles and norms established in regimes (see Table 9.1). Principles and norms which are explicitly referred to by legal documents were identified for 126 regime elements. Table 9.1 illustrates that 765 data are included in this data-set describing basic principles and norms which are explicitly (542) or implicitly (223) mentioned in treaties that constitute a regime. First of all, analysis will focus on those principles or norms which are explicitly mentioned in legal documents. A significant amount of these data refers to *basic norms of international society* that govern the interactions among states in international institutions. Norms like the "sovereignty of the nation-state" or of "sovereignty over natural resources" have been coded in 72 and 70 regime elements. The two sovereignty norms have both been frequently established in a regime element. Accordingly, a subset of 54 regime elements includes each of the two norms. For 82 regime elements, the norm of "good neighborliness and international cooperation" has been established. *Basic environmental principles and norms* dominated the legal meta-structure of regimes. The "precautionary principle", the "polluter pays-principle", the norm prescribing "responsibility not to cause damage to the environment", the principles which demand "preventive action" or emphasize the "common concern" about the

Table 9.1   Important Regime-Specific Principles and Norms in Regime Elements

Principles and Norms in Regime Elements	Total	1 = Constitutive Agreements Make Explicit Reference to Principle/Norm	2 = Commitment to Principle/Norm is Implicit
1 = Precautionary Principle	63	43	20
2 = Polluter Pays	46	36	10
3 = Sovereignty	72	33	39
4 = Sovereignty Over Natural Resources	70	39	31
5 = Responsibility Not to Cause Damage to Environment	80	64	16
6 = Principle of Preventive Action	93	77	16
7 = Good Neighborliness and International Cooperation:	82	84	8
8 = Common Concern	76	61	15
9 = Common Heritage of Mankind	26	4	22
10 = Intergenerational Equity	26	16	10
11 = Equitable Use of Natural Resources	21	9	12
12 = Sustainable Use of Natural Resources	57	47	10
13 = Integration of Environment and Development	40	28	12
Total	765	542	223

consequences of an environmental problem were coded in 281 instances altogether. Mostly more than one of these environmental norms applied to a regime element. In 115 regime elements belonging to 19 regimes explicit reference has been made to one or more environmental principle or norm.

Less than a fifth of the total consists of data that describe the existence of explicit *basic principles and norms of equity and sustainability*. Sixteen of the 21 regimes include at least one of these norms of justice. The "principle of sustainable use of natural resources" has most frequently been referred to for 47 regime elements that belong to 13 regimes. It is followed by the "principle of integration of environment and development" which is included in 28 regime elements belonging to eight regimes. The "principle of intergenerational equity" has explicitly been referred to in treaties that constitute 16 regime elements in five regimes. In contrast, only within two regimes (or four regime elements) an explicit commitment has been made by members to the "common heritage-approach". The "equitable use of natural resources" has explicitly been identified as a principle in three regimes (or nine regime elements). In some regimes, principles and norms of justice were considered as so important that coding experts selected them as important goals. For example, the goal of an equitable sharing of benefits arising out of genetic resources became established in the biodiversity regime.

The existence of additional implicit commitments to various principles and norms of justice has been detected for 10 regimes. This suggests the conclusion that an ongoing development of norms of justice has been occurring in many regimes. The comments of coding experts also illustrate that international environmental treaties have often been concluded before international norms of justice were developed.

Beside the existence of explicit norms of justice in the Ramsar regime on wetlands, the commitment to various other norms is implicit because the "Ramsar Convention was adopted before certain principles were articulated or formalised at the international level". Such principles and norms "inform many more recent resolutions and recommendations adopted by the Ramsar Conference of the Parties".[4] This points to the fact that global discourse about distributional justice became even more established in soft-law documents than in the legal provisions of regimes. The analysis of those principles and norms of justice that are explicitly provided in international legal agreements also justifies the conclusion that distributional justice has been acknowledged as an important issue by states in constitutive legal agreements of the broad majority of regimes that were coded for the database.

Distributional Consequences

Costs and benefits can be understood in a broad sense. Material or non-material costs are those which result from institutional provisions that demand the implementation of reduction measures for pollutants, of measures for the conservation of single species, or of other actions required to fulfill the obligations

---

4   Comment on the coding of this variable made by Dwight Peck and Clare Shine.

of a regime. The benefits which can be gained from the operation of a regime pertain to possible improvements in the state of the environment or can include gains affecting rights which govern the use of a resource. Other types of costs and benefits are perceivable which occur mainly outside the issue-area. Indirect costs and benefits can exist for which it is often difficult to determine in how far they were caused by a regime. The analysis will focus on regime consequences that occurred within the environmental issue-areas that were linked with the operation of a regime. Attention will primarily be paid to the direct impacts of a regime.

Information about the distribution of material or non-material costs and benefits can be derived from written narratives provided for 22 regimes. These narratives suggest to distinguish four types of regime consequences: i) the distribution of the costs occurring for the administration or management of a regime; ii) economic costs and benefits arising in conjunction with the implementation of regime policies; iii) environmental benefits; iv) political costs and benefits that occur from the existence of a regime. It will be explored whether these impacts reflect differentiated responsibilities of states for the causation of a trans-boundary problem and whether different levels of socio-economic development were considered in the development of regime policies. From a normative perspective, the existence of strong differences in these levels of socio-economic development between industrialized and developing countries means that prosperous countries should bear a proportionally higher share of the costs than countries with lower economic, financial, or technological capacities.

Costs for the Administration and Management of Regimes   In most of the regimes procedures had been developed which governed the division of costs for the administration and management of regimes. In a significant number of cases, costs were split among the Contracting Parties according to the UN scale of assessment of GNP or they were based on similar calculations. This procedure is used, for example in the ozone regime for the collection of member contributions, and as a result countries "that would according to the (adjusted) UN scale of assessments have to contribute minor amounts (less than 0.1 per cent) do not need to contribute".[5] This uneven distribution of state contributions to regime budgets considers different levels of socio-economic development. The cost-sharing formula agreed upon in the context of the Convention on the Rhine against Pollution for the financing of the secretariat of the Rhine regime imposed a similar amount of the costs to Germany, France and the Netherlands (28 per cent each), whereas the Swiss Confederation (14 per cent) and the Grand Duchy Luxembourg (2 per cent) carried the rest of these costs. This formula was only slightly changed over the years as the European Union began to contribute a minor portion to the annual budget.[6] Considering that Germany and France both exceed the Netherlands

---

5   Written narrative about the distribution of costs made by Sebastian Oberthür.

6   For the formula that defines the sharing of costs for the annual budget between Germany, France, Luxembourg, the Netherlands, the Swiss Confederation and the European

with respect to the size of their national populations or their GNPs, this formula involves that the Netherlands as a downstream country bears a proportionally higher financial burden than other countries.

Member contributions to international fisheries organizations were, either in parts or fully, calculated with reference to the user pays principle. They were based on different formulas which charged general membership fees or raised revenues by the calculation of levies that considered a country's proportion to catches. On the one hand, this form of the calculation of fees seems justified as it considers the proportional gains made by harvesters of a resource. On the other hand, such formulas neglect to consider inequalities that exist with respect to different levels of socio-economic development among member countries. The early period of the international whaling regime was financed by contributions calculated as a flat-rate levy on all the members, but from the early 1960s "there was a levy placed on the Antarctic whaling states to pay for the special investigation into the whale stocks". The decline of the number of active whaling nations involved that "a new formula for calculating contributions was introduced which in addition to a flat-rate component had additional shares for engaging in whaling activity". A third element for share contribution which considered the size of the delegation attending the annual meeting was added later on.[7] The amount "of payment of member country contributions into the ICCAT-budget (...) is determined by a formula based largely on the amount of ICCAT-fish caught and canned by each country".[8] Against this background, it has been concluded that "the states with larger catches and high consumption pay much higher contributions to the costs of running the commission".[9]

The amount of total costs varies among the regimes. In some regimes, the amount of costs that occurred on the international level was insignificant. For example, matters of the Barents Sea fisheries regime were predominantly carried out on the national level and no regime secretariat was established until the end of the exploration period. A similar form of cost-sharing emerged between the United States and Canada in the context of the Great Lakes management regime where each of the two offices of the International Joint Commission was financed by that government on whose territory it was located. Research activities related to the environment of the Great Lakes were primarily funded by each nation through related agency budgets. The regime that emerged with the Convention on the Prevention of Marine Pollution by Dumping of Wastes and Other Matter (London Convention) did not have a regular budget of its own until the end of the exploration period. But IMO has been responsible for the administration of regime-related matters.

Union see Article 9 of the Rules of Procedure and Financial Regulations of the ICPR as last amended by the 67th Plenary Assembly of the ICPR, 3rd July 2001 in Luxemburg.

7   Written narrative about the distribution of costs made by Ray Gambell.
8   Written narrative about the distribution of costs made by Brian Hallman.
9   Written narrative about the distribution of costs made by James S. Beckett.

The lack of a regime secretariat for the Antarctic regime implied that the "administrative costs of an ATCP-meeting is borne by the host government of a state where the meeting is held". A separate budget exists for the management of the Commission on Antarctic Marine Living Resources. The Convention on the Conservation of Antarctic Marine Living Resources provides that member contributions shall be determined on the basis of a formula that considers the amount of resources harvested by parties and an equal sharing among all members of the Commission [Article 19(3)]. Additional funds for the operation of the Convention for the Conservation of Antarctic Seals or for CCAMLR were provided by national governments.[10] Some Antarctic Treaty Consultative Parties (ATCPs) "pay considerably more for their activities in the Antarctic than other governments" because each government "supports its own scientific/research activities in the region". Under the circumstances, "higher absolute costs were borne by those states maintaining the most active research programs and/or enforcement efforts" in the regime.[11] Many of these efforts were primarily carried out to achieve the goals of the Antarctic regime, but they also contributed to secure the political claims or security interests of these states in the Antarctic region. Evidence for voluntary contributions of single countries exists for a broad number of regimes. Additional financial support was frequently provided by the host country of a regime secretariat. Voluntary contributions predominantly came from industrialized countries which intended to finance ad hoc workshops and to support the work of programmatic activities. To sum up, in a significant number of cases the financing of the administration and management of regimes considered different levels of socio-economic development. However, formulas which calculated fees on the basis of a country's proportion to catches occasionally failed to consider the asymmetries that exist with respect to the financial capacities between industrialized and developing countries.

Implementation Costs and Benefits   These costs can pertain to the investments which must be made by states, societies or economic actors for implementing international policies on the domestic and transnational level. They also include those costs which occur from the restriction of social practices arising from regime policies (e.g., limitations on the exploitation of resources) or which arise in connection with the enforcement of international policies. In some respects, an assessment with regard to whether implementation costs or benefits reflect equitable solutions is contingent on characteristics of the type of good management by a regime. For example, the control of trans-boundary externalities refers to problems where one country is the victim of another country's externalization of a certain problem. Higher costs for the implementation of measures which reduce the trans-boundary export of hazardous waste or river pollution seem justified

---

10   Written narrative about the distribution of costs made by Christopher C. Joyner.

11   Narratives on the distribution of costs made by Christopher Joyner and MJ Peterson.

for those countries which mainly export their pollution to other countries. The damaging effects that occur from the overuse of common pool or common property resources suggests that resource units (e.g., an acre-foot of water for a ground water basin, tons of fish for fishing grounds) or pollution rights must be established that reflect normative considerations of equity or sustainability. However, uneven interests between major causers and those which are mainly affected by a problem involves that a reallocation of these rights which is necessary for reducing the pressure on ecosystems is also determined by considerations other than equity. A reassignment of these rights can often not fully equal the ideal solution that would emerge if reference would exclusively be made to normative considerations, but regime members intend to realize their own interests in institutional bargaining. The following description is based on generalized statements of coding experts which assessed the division of costs and benefits across the board. Implementation costs must not inevitably occur for each regime member. For example, non-whaling states suffered no implementation costs from the whaling regime but experienced political benefits because the regime contributed to fulfill their goal of the conservation of whale stocks. In contrast, the introduction of the ban on whaling involved that the whaling nations suffered a loss of products and income (Zangl 1999, 260–83).

Socio-economic and geographical factors like the length of national coasts or riversides determined the amount of costs which individual countries had to bear for cleaning up regional seas or trans-boundary rivers. Each member state of the Rhine regime has the same relative burden since each state is responsible to finance the measures needed on its own territory. For the regime component which deals with the reduction of chloride pollution, a cost-sharing formula has been agreed upon between the four riparian states that "reflects the relative contributions of the individual countries to the pollution problem and the intensity of their demand for chloride reductions" (Bernauer 1996, 210). Downstream countries financed measures for the reduction of chloride pollution that were implemented on French territory. These measures caused environmental improvements in downstream countries. Differences with respect to the distribution of costs for the implementation of MARPOL-provisions were largely dependent on the size of national fleets. Flag states "incurred the largest costs due to the cost of regulating tankers. Port states incurred costs for inspecting ships and for ensuring provision of reception facilities. Coastal states (if lacking national fleets or huge ports) incurred few, if any, costs".[12]

Considering that a significant contribution to marine pollution stems from land-based sources, it seems plausible that states with long coasts, with densely populated areas along the coastline, or with hot spots that contribute to marine pollution carry higher costs than those countries which play a less important role as a causer of the problem. The conservation of wetlands and the management of water issues addressed by the Ramsar regime on wetlands was combined with

---

12   Written narrative about the distribution of costs made by Ronald B. Mitchell.

implementation costs that broadly varied according to the scale of the problem and the biophysical characteristics of each country. Under the circumstances, some "industrialised and/or densely populated countries, such as the Netherlands, faced high costs for conservation of remaining areas and restoration of wetlands".[13] A country's proportionally high share of implementation costs must not only contribute to improving the state of the environment beyond national borders. Measures which are implemented within a country's own territory can also improve the inland environment. It is frequently impossible to assess costs mainly in terms of relative gains among states, but a cost-benefit-ratio includes also those costs and benefits that occur on the domestic level.

Limitations imposed on the catch of tuna or other living resources in conjunction with other conservation measures involved that some states suffered higher costs in some resource regimes than others. For example, the conservation and management of dolphins which has been realized as an important task in the IATTC-regime since the mid 1970s involved that "costs to states that relied heavily on fishing tuna in association with dolphins, such as Mexico and Venezuela, were great because of restrictions placed by the regime on such fishing and because of embargoes applied by the USA against fishing tuna with dolphins".[14] Similar impacts occurred occasionally for those distant water fishing nations whose claim to the use of regional fish stocks was considered less legitimate. In contrast, costs and benefits in the Barents Sea fisheries regime were more or less symmetrically distributed among Norway and Russia which had "50 per cent of the cod quota each and would suffer the same economic loss in case of mismanagement".[15] In some resource regimes the cost-benefit-ratio seems to have changed in the longer term. In particular, the slowing or reversal of resource reduction involved that the pressure on tuna fish-stocks could be reduced in both tuna regimes.[16] Thus, states which suffered high cost from the implementation of conservation measures in these regimes in the short term contributed to avoid higher costs for themselves and for other fishing-nations that would have possibly arisen from continued depletion of fish-stocks. The need for maintaining fish-stocks in particular regions at sustainable levels which will secure that these stocks will also be available to future generations justifies that stronger limitations will be imposed on catch-quota of distant water fishing nations than on regional fishing-nations.

Differences in the costs that occurred in connection with implementation were sometimes necessary for the balancing of different levels of socio-economic development. Occasionally, these differences also reflected differentiated responsibilities for the causation of a problem in an issue-area. Such differentiated responsibilities for the causation and management of a problem were accepted by industrialized countries in regimes addressing global issues like climate change or

---

13   Written narrative about the distribution of costs made by Clare Shine.
14   Written narrative about the distribution of costs made by James Joseph.
15   Written narrative about the distribution of costs made by Christel Elvestad.
16   Written narrative about the distribution of costs made by James S. Beckett.

ozone depletion. In a number of regimes countries with economies in transition were given financial support in order to secure their cooperation. Various regimes were determined by cost-benefit-relations that favored developing countries vis-à-vis industrialized countries. For example, agreed incremental costs occurring as a result of the implementation of the biodiversity convention or of other global environmental agreements are borne by developed countries. Developed countries, "whether they are experiencing desertification or not, are to contribute the bulk of financial, scientific and technological resources for the operation" of the desertification regime.[17] Regulations for the stabilization of the emissions of greenhouse gases that emerged in the context of the Kyoto-Protocol specified that industrialized countries shall – compared to the level of their emissions in 1990 – achieve a reduction of about 5 per cent during 2008 and 2012, whereas developing countries remained unaffected from such reduction measures (Oberthür and Ott 2000). Costs and benefits combined with the implementation of reduction measures for greenhouse gases or with emission trading will occur predominantly beyond the end of our exploration period during the first decade of the new century.

Developing countries were given a grace period before they had to begin to stabilize, reduce, or finally phase-out the production and consumption of ozone depleting substances (Parson and Greene 1995). It could be argued that industrialized countries faced higher costs than developing countries in this regime, but their industries were also able to "market alternative substances/technologies and sometimes gain special benefits from those alternative options (e.g., reduced energy consumption etc.)". Furthermore, it could be argued that differences regarding the distribution of economic costs and benefits also occurred between industrialized countries. Those countries which had begun to phase-out the consumption of ozone-depleting substances in specific sectors (e.g., the U.S. in the aerosol sector) prior to the temporal baselines that were included in the Montreal Protocol and in subsequent legal arrangements could no longer use this cheap reduction potential to achieve compliance with regime rules. The European Union which had avoided the implementation of strong reduction measures until it finally agreed to the Montreal Protocol could use this reduction potential for complying with regime rules. Nevertheless, this case also illustrates that uneven starting-positions for the reduction of pollutants must not inevitably lead to an asymmetrical distribution of costs and benefits in the long term. Rather, it has been observed that the early implementation of reduction measures involved "that the state of development of substitute chemicals appears to have been particularly advanced in the U.S. at the end of the 1980s and the U.S. industry has gained a much larger share of the global HCFC substitute market than it had in the global CFC market".[18] This example illustrates that international regimes are not only limiting the activities of economic actors, but also provide new market-conditions that can be used by firms as new profitable opportunities. On balance, costs and benefits arising from

---

17   Written narrative about the distribution of costs made by Elisabeth Corell.
18   Written narrative about benefits made by Sebastian Oberthür.

the implementation of policies were unevenly distributed in many cases. Though complete information about the magnitude of these costs may not always have been available at the time when states agreed to implement international policies, they mostly understood that geographical factors like long coasts and riversides or the proportionally high contribution of their national industries to transnational pollution could cause considerable costs during implementation. In the long term, benefits occurred also for many of those countries which originally suffered the major financial and economic burden – whether these benefits were economic, environmental, or political.

Environmental Benefits   Assessments regarding costs and benefits can turn out differently depending on whether environmental or economic impacts will be considered. It has been argued with regard to the impacts of the CITES-regime that conservation effects of the regime can be considered as global benefits and that the regime "effectively reduced the initial competitive advantage of free-riders, by broadening global coverage and by tightening enforcement during the second phase" in the 1990s.[19] In contrast, it could be argued that this way of assessment reflects a conservationist view that partly ignores those economic costs on sub-national levels that were combined with these benefits in the short term. Under the circumstances, it has been argued by another expert that "while at the highest level of abstraction it might be possible to say that all people benefit from a protected ecosystem around the globe, this is too tenuous". Rather, it has been established that "as the number of species in need of protection vary so do the ecological benefits" and that "there can be considerable economic loss, if not at the state level at least at the local level when trade is restricted".[20]

The evolution of environmental or other benefits frequently depended on the type of problem managed by a regime. The environmental benefits resulting from the ozone regime are deemed "to be distributed asymmetrically since the avoided ozone loss is predicted to increase towards the poles" and thus "temperate regions (where most industrialized countries are located) should have benefited more than tropical regions (where most developing countries are located)".[21] Water quality of rivers in downstream countries particularly benefited from the abatement of pollution in upstream countries. Impacts of regimes for the protection of the Rhine or the Danube illustrate that each riparian country of these rivers experienced environmental benefits – although they were not always evenly distributed. Obviously, the distribution of benefits must not inevitably remain stable in the long-term. The LRTAP-regime produced benefits in its early period that were relatively concentrated in those countries which had originally been importers of pollution, whereas a more even distribution of benefits emerged in the long term. Conservation measures taken in resource regimes led to long-term benefits for

---

19   Written narrative about costs made by Peter H. Sand.
20   Written narrative about costs made by David S. Favre.
21   Written narrative about the distribution of costs made by Sebastian Oberthür.

each regime member. It has been established as an essential benefit resulting from the ICCAT-regime that "there will be healthy and productive highly migratory fish stocks in the Atlantic ocean" – whether they occur within the 200 mile zone or on the high seas. In addition, countries with the biggest fishing industries associated with the ICCAT-stocks and the ones with the most fish off their coasts seemed to have experienced particular benefits.[22] Similar benefits were identified for the IATTC-regime "because resources were maintained at high levels of abundance" and "all states that exploited these resources benefited from higher catches resulting from greater abundance of tuna".[23]

Even so, environmental benefits arising from the desertification regime will mostly occur in developing countries. Similar benefits are likely to arise from the Basel Convention for developing countries. It has been observed that countries which "otherwise might unintentionally or not be able to prevent hazardous waste imports receive more benefits than states that would otherwise wish to export hazardous wastes without international constraints. Moreover, in the case of an illegal or harmful shipment of hazardous waste arriving in a developing country, they will receive assistance that would not be provided to an industrialized country".[24] Various examples illustrate that industrialized countries strengthened their efforts to improve the state of the environment in developing countries, even though this engagement may still not be sufficient in individual regimes. Industrialized countries accepted that they are partly responsible for the deterioration of the environment and for the depletion of natural resources in developing countries. It is often difficult to determine how far the behavior of industrialized countries in regimes is based on their belief in the relevance of the principles of equity and sustainability. It could be argued that industrialized countries provide these incentives in order to fulfill their own interests in environmental regimes when participation by developing countries is required to avoid free-riding or other negative impacts. But this argument is not applicable when industrialized countries contribute material benefits without receiving a significant direct environmental or economic benefit on their own territories.

Political Costs and Benefits These regime consequences can occur in many different ways. International negotiations can produce political costs and benefits on the domestic level that can affect the prospects of national governments in elections. The policies and programmatic activities of regimes can involve that particular non-state actors experience a strengthening or loss of their influence in the issue-area. Political costs and benefits that are relevant in terms of the distribution of values were primarily identified for the inter-state level. The Antarctic regime can be used for illustrating that countries with a special political interest in a regime have to bear higher material costs than others for securing their political

---

22 Written narrative about the distribution of benefits made by Brian Hallman.
23 Written narrative about the distribution of costs made by James Joseph.
24 Written narrative about cost-benefit-relation made by Jonathan Krueger.

claims. The United States, the United Kingdom, Australia or other claimants paid considerable more for their activities in the Antarctic than other governments had to pay for their research activities, for the maintenance of various stations, or for their engagement in single components of this regime. These special activities also served to maintain their influential role in this regime. Non-claimant states benefited from improved access to the Antarctic regime. Under the circumstances, the question arises on whether the expansion of membership led to higher costs for these newly acceding members for maintaining status as a Consultative Party "because they must maintain activity to keep status while the 12 original parties (7 claimants, 5 non-claimants) retain status regardless of later level of activity by virtue of being an original party".[25] The uneven treatment of claimants and non-claimants has been a contentious issue for several decades, but greater access of non-claimants to the continent and to decision-making of the Antarctic regime reduced, though not completely abolished, these differences (Vidas 1996, 55–7). On balance, all members have benefited from a reduction of conflict about territorial claims in Antarctica. Similar political benefits emerged from the Protocol on Environmental Protection, since conflict about the exploitation of resources could be reduced. The South Pacific fisheries regime can be used as another example for illustrating that individual states provide a significant amount of material resources in order to secure their own political influence. In this context, it has been argued that "Australia and New Zealand pay the greatest direct financial costs by far but neither derives anything like a proportionate direct benefit from the regime at least in terms of its formal objectives". It was also pointed out that "if diplomatic advantage and direct access to the region are counted as benefits, however indirect, perhaps the two countries get a fair return for their contribution to the FFA".[26]

Summary

The codification of principles or norms of justice in constitutive agreements of regimes indicates that global governance systems could not avoid responding to these demands that were raised in transnational discourse. Developing countries can refer to these norms in negotiations with industrialized countries for achieving additional support for the implementation of international policies and for improving those social living conditions that contribute to damage the environment in the first place. In many regimes we still lack a clear understanding on how these norms of justice can be operationalized and measured. The findings concerning the costs for the administration and management of regimes suggest that industrialized countries often bear the major financial burden in those regimes where membership is composed of both industrialized and developing countries. The procedures chosen for the calculation of fees in some resource regimes

---

25   Written narrative about the distribution of costs made by MJ Peterson.
26   Written narrative about cost-benefit-relation made by Richard A. Herr.

raise the question with regard to whether they take different socio-economic capacities of members into account adequately Implementation costs can broadly vary among regime members, but mostly socio-economic, geographical or other factors (e.g., number of hot spots, geographical characteristics) account for these differences. The practice to externalize pollution or costs for abatement measures seems no longer legitimate, but the evolution of environmental principles or norms of justice indicates that responsibilities for the causation of a problem or for the equitable distribution of costs and benefits are expressed in constitutive agreements of regimes. While societies begin to realize that growing depth of cooperation increasingly affects economic wealth on sub-national and local levels, they may also expect that norms of justice and sustainability will be expressed in regime policies and not remain rhetoric commitments.

Institutional Mechanisms and the Distributional Consequences of Regimes

How far did institutional mechanisms contribute to change distributional impacts of regimes? Did a growing consideration of normative claims for equity and sustainability through the principles and norms of regimes find expression in the creation of institutional mechanisms that can reduce those uneven starting-positions of states for implementation which emerge from different levels of socio-economic development? The fact that qualitative statements were used for determining distributional consequences in regimes involves that direct correlation between changes in these impacts and the role of institutional mechanisms can only be identified for certain regimes. Data which describe the existence of programmatic activities like financial and technology transfer or of financial mechanisms like trust funds can be used for exploring whether these institutional characteristics had an impact on distributional consequences of regimes. The establishment of funding mechanisms for financial and technology transfer became predominantly necessary in regimes which combine members from industrialized countries and from the developing world (Streck 2006). Accordingly, findings about the influence of funding mechanisms on distributional consequences primarily pertain to the subset of regimes where the management of conflicts between North and South was necessary for problem-solving. Measures for the funding of financial and technology transfers on the level of a regime were implemented in about half of our total number of regimes. Differences existed among these mechanisms regarding the amount of resources made available for the financing of projects or with respect to the scope of their activities. Donor states could normally prevent that regimes would be given their own distinct source of revenues, but funds that became available came mostly from national contributions and assessments on members. While emphasis will be placed on those contributions of funding mechanisms to impacts on the domestic level of developing countries, it is worth mentioning that another important task of trust funds involved to strengthen participation of developing countries in meetings and programmatic activities of regimes. These mechanisms contributed in some regimes that developing countries or countries with economies in transition which could be affected from the impact of regimes could also participate in regime discourse and decision-making. A case in point is the establishment of a special voluntary fund to support participation of developing countries in the climate change negotiations. The provision of these funds made it possible that travel and subsistence costs could be offered "to one delegate each from 99 developing countries (including all the least developed countries and many small island developing countries)" to ensure participation in the second session of the Intergovernmental Negotiating Committee for a Framework Convention on Climate Change in June 1991. Consequently, "the number of delegations from developing countries at this session with representation from capitals had more than doubled by comparison with the first session".[27] Developing countries also took advantage of these funds in subsequent sessions.

International trust funds or similar mechanisms financed incremental costs for developing countries to meet their obligations in various international regimes or enabled industrialized countries to fulfill their obligations partly by the financing of measures in countries from the developing world or in countries with economies in transition. The Global Environment Facility (GEF), implementing agencies and other international organizations coordinated and managed the implementation of financial and technology transfers in several regimes. While the GEF is the designated financial mechanism of the biodiversity regime and the climate change regime, it also granted funds for the implementation of measures that contribute to phase out the use of ozone depleting substances. The implementation of these measures is coordinated with various bodies that were established in these regimes. GEF had allocated around a billion dollars for the funding of 345 biodiversity projects until the beginning of the year 2000. Additional matching funds provided by the co-financing of other donors more than doubled the amount of funds that became available for these projects. Under the circumstances, it has been argued that the conservation costs arising in the context of the biodiversity convention were initially borne by developing countries but they were "then transferred to developed countries as they committed themselves to provide funding for incremental costs".[28] Industrialized countries in the climate change regime committed to provide new and additional financial resources and to facilitate technology transfer to developing countries. Various funds were established in the context of the Climate Change Convention and the Kyoto Protocol for the financing of costs which occur for so-called non-Annex I countries (e.g., developing countries) in conjunction with the submission of their national inventories, to finance projects for the reduction of emissions of greenhouse gases in developing countries, to promote the transfer of technology and to stimulate the use of less energy-intensive technologies in developing countries, and to finance projects for the adaptation to climate change – to mention only a few of the tasks

27   See United Nations General Assembly A/AC.237/9, 20.
28   Written narrative about funding mechanisms made by Gudrun Henne.

managed by these funds. While these institutional mechanisms can be deemed to produce positive effects in developing countries, it is also "well-known that reliance on consultancy and technologies sourced in industrialized countries often leads to benefits flowing back to donor countries".[29]

The creation of international funding mechanisms demonstrates that development aid policies of industrialized countries were increasingly linked to the goals of international environmental agreements. In many environmental regimes the funding of developing countries bears the character of green development aid through which industrialized countries intend to improve the socio-economic conditions in these countries just as much as needed to fulfill their own environmental goals. Self-interest must not inevitably be a dominating factor for the readiness of industrialized countries to engage in financial and technology transfers with developing countries. It has been observed for the South Pacific fisheries regime that it is "treated by its developed members (Australia and New Zealand) and by many other states [both distant water fishing nations and other aid donors] as a development activity" where many of the "costs are carried by states who are not the primary beneficiaries and in some cases are not even directly affected by the regime".[30]

There is another example which can illustrate that only the provision of institutional mechanisms for financial and technology transfer improved the distributional impacts of a regime for developing countries. During its early period until the mid 1980s, mainly countries from the developed world had joined the Ramsar-regime on wetlands. In the regime's second period from the late 1980s onward, the distribution of benefits became less symmetrical as a result of growing attention paid to the conservation and sustainable use of wetlands in developing countries. It has been argued that in developed countries, direct benefits of a member state resulted from assistance of NGOs and of courts which protected wetlands from unwise development, "whereas in the developing world, a far greater range of benefits can be enjoyed, ranging from technical and financial assistance (either from or in the name of the Convention) to guidelines and best-practice information to assist them in dealing with their problems more successfully". Under the circumstances, it was concluded that "it is fair to say that presently the benefits of joining the Convention are substantially different for developed and developing countries".[31]

It should be noted that in addition to the environmental aid that was provided on the level of various regimes, individual countries or external organizations frequently initiated the funding of projects with countries from the developing world or in countries with economies in transition. The CITES-regime has been determined by deep conflict between states which were wishing to protect species (e.g., the elephant) and range states where people on the local level would have

29   Written narrative about funding mechanisms made by Jacob Werksman.
30   Written narrative about funding mechanisms made by Richard A. Herr.
31   Written narrative about funding mechanisms made by Dwight Peck.

suffered economic costs from conservation measures. The United States and other states "have established funds in the 1990s to provide direct financial support for some range state protections to help with the economic cost of protection".[32] After the United States Congress had enacted the Elephant Conservation Act in 1988, 12 million dollars had been allocated to the African Elephant Conservation Fund which is managed by the US Fish and Wildlife Service "resulting in the awarding of 170 grants for projects in 25 range countries leveraging 57 million dollars in matching and in-kind support for the protection, conservation and management of African elephants" (US Fish & Wildlife Service 2003). In 1994, the United States Congress passed another act which established the Rhinoceros and Tiger Conservation Fund for the provision of resources to support conservation activities for these species in Asia and Africa.

So far, the provision of financial and technological aid has been discussed as an issue that becomes predominantly relevant in the North-South-context. A rather untypical case for such a transfer between industrialized countries has been the significant contribution made by Germany, the Netherlands and Switzerland to financing reductions of chloride emissions of a state-owned potash mine on French territory. Two projects had been launched for achieving these reductions at the end of the 1980s and in the early 1990s. Only about a third of the costs for these measures were carried by France. The major portion of costs was transferred to France by the three other riparian countries in form of lump-sum payments, whereas a joint fund has not been established. Uneven concern in the different countries and asymmetries between the Netherlands as an affected downstream country and France as a polluting upstream country made it difficult for these countries to arrive at a solution other than to finance reduction measures on French territory, although the technological and financial capacities of France went far beyond the level of a developing country. This transfer may have contributed to reducing chloride pollution of the Rhine in affected downstream countries. From a normative perspective, it contradicts with the special responsibility combined with the polluter pays principle that a polluting country has to bear a proportionate share of the costs. Thus, this type of financial mechanism reflects an uneven distributional outcome if responsibilities for the causation of a problem, the capacities available for problem-solving, or the proportionate share of the costs occurring for riparian states for problem-solving will be considered (Bernauer 1996).

These findings suggest that growing consideration of principles and norms of justice in constitutive legal agreements found expression in the creation or expansion of institutional mechanisms which support developing countries in the implementation of international environmental policies. Whether these funds are always sufficient to compensate these countries for the costs incurring in connection with the implementation of international policies is part of on-going debate within many regimes.

---

32 Written narrative about funding mechanisms made by David S. Favre.

Non-State Actors and the Distributional Consequences of Regimes

While distributional consequences of regimes were identified on the basis of written narratives, the impact of non-state actors will also predominantly be illustrated on the basis of qualitative studies. The contributions made by non-state actors to the evolution of principles or norms of justice or to changes in distributional consequences will be illustrated for individual regimes. But it is impossible to make an assessment which can be based on a quantitative data-set. Growing consideration of concerns for equity by the principles and norms of regimes partly emerged from discourse in international society regarding the structural inequalities of the global economy which is biased against the interests of developing countries in many respects. The ideological basis for these principles has been established by efforts of the "Group of 77" which argued for a reduction of asymmetries in trade and development between North and South. This debate has been reframed in such a way that developing countries made their willingness to accede to and implement international environmental agreements conditional upon the consideration of aspects of equity in these regimes (Biermann 1998, 45f). The UN world conferences on the environment held in Stockholm (1972) and in Rio de Janeiro (1992) were both events that stimulated the development of norms of equity and sustainability. In principle, industrialized countries acknowledged in Agenda 21 that these norms should become constitutive elements for the governance of international environmental issues. But skeptics also raised doubts whether this moral approach to international environmental law will yield fruit. It is certainly difficult to operationalize some of these norms, but their existence can support the relevance of the arguments made by developing countries in negotiations with industrialized countries. Three different ways through which non-state actors supported the development of norms or contributed to change distributional consequences in certain regimes will be illustrated in the following. It will become obvious that these actors engaged in discourse about the evolution of norms of justice, that they supported developing countries in their effort to call attention to the need for the establishment of financial mechanisms in regimes for capacity-building, and that problem-solving in regimes also benefited from additional financial support or expertise provided by private actors.

At first, international legal scholars became increasingly engaged in the development of legal principles and norms of equity or sustainability. Activist environmental groups referred to these norms and supported their relevance in public statements. These activities affected norm-evolution in the broader field of global environmental law just as much than in single regimes. For example, IUCN established an environmental law program which consists of a global volunteer network of several hundreds of environmental law experts. They assist decision-makers with various kinds of legal services required in connection with international environmental negotiations. The establishment of international norms of sustainability and equity in environmental regimes emerged partly as a result of the recognition of these norms in global reports developed by the World

Commission on Environment and Development or in soft law documents such as the World Charter for Nature or Agenda 21.

Second, the question arises on whether activities of non-state actors could contribute to changing the distribution of values by regimes in such a way that they appeared more appropriate than in terms of normative considerations. This question is complex and will require more exhaustive studies. One could argue that these actors drew attention to those impacts of regimes that are not equitable or sustainable. They also supported the idea that new institutional mechanisms will have to be established for the production of more equitable and sustainable solutions in regimes. Negotiations in regimes dealing with climate change, biodiversity, or ozone depletion can be referred to as examples where non-state actors supported the political claims of developing countries for the establishment of financial and technology transfers. The operation of programmatic activities in regimes can provide a starting-point for these actors to refer to asymmetries between North and South. For example, it has been observed that the climate regime's "procedural requirement to review and adjust commitments have maintained climate change as a central element of North-South politics as well as trans-Atlantic tensions" and that the "attention the regime consistently draws to disparities in energy efficiency and consumption patterns has raised the profile of debates on equity and sustainable development".[33] It is also obvious that non-state actors can be one of those forces in world politics which link discourses between related topics so that concerns for equity or sustainability will spill over in other issue-areas. It has been observed that the biodiversity regime stimulated "the inclusion of the protection of the knowledge, innovations and practices of indigenous and local communities embodying traditional lifestyles relevant for the conservation and sustainable use of biodiversity in WIPO and to some extent WTO", and non-state actors made efforts to raise the possible impacts of the policies of these institutions as an issue. Non-state actors called upon industrialized countries to provide more financial or other resources which can be used for compensating developing countries for the costs incurring from the conservation of species or forests. But they also criticized some countries of the South for environmentally damaging practices, for tolerating deforestation, or for the depletion of protected species.

Third, participation of non-state actors in the implementation of environmental policies in developing countries is another factor that can contribute to improve distributional consequences emerging for these countries from a regime. Many of these services of non-state actors for capacity-building were illustrated in previous chapters. Such activities of non-state actors can also be carried out independently from a regime. Private sponsorship can be used as an instrument by the emerging global civil society for harmonizing social needs in developing countries with environmental protection. For example, projects of WWF aim at improving the

---

33   Written narrative made by Jacob Werksman about the changes that occurred in the contents of the international political agenda or the priority of issues discussed in this regime.

protection of ecosystems or conservation of species in developing countries. The eco-development projects of WWF carried out in the past were aimed to protect the environment at the local level and thus bridged the distance that exists between international environmental institutions and local living conditions (Wapner 1996, 72f). In the 1990s, multinational firms began to support eco-development projects of non-profit-oriented service-organizations in developing countries through the provision of funds or expertise. It is hard to judge whether these funds were predominantly allocated by these firms in order to improve their public image or because of ethical motivations of top managements. In some respects, there is evidence that both reasons stimulated the engagement of private firms in these projects. The establishment of the Global Compact between the United Nations and multinational firms took the relevance of these actors for global policy-making into account (Paul 2001, 122–8). Less formalized agreements or codes of conduct agreed upon between private actors or between private actors and states can be considered as supplements to international regimes.

Conclusion

The international society accepted that the evolution of social order can only be achieved if international regimes produce impacts which are considered as equitable by affected states. Non-state actors have not been the only driving forces which supported the evolution of norms of justice or which politicized distributional consequences of regimes. Rather initiatives of developing countries influenced developments in regimes in many respects. Industrialized countries have accepted their special responsibilities for the provision of financial and technological assistance to developing countries. For the time being, the conflict between North and South about the amount of resources needed for managing trans-boundary problems effectively can be assumed to continue, partly because the management of environmental issues is closely linked to improving social and economic living conditions in developing countries. Institutional mechanisms for financial or technology transfer can contribute to improve the capacities for implementation in the developing world. Bilateral assistance or improved consideration of environmental aspects in development aid policies of industrialized countries may serve as additional instruments. The rising relevance of private or semi-private forms of governance in world politics suggests that the traditional division of labor between states and private actors for the management of transnational problems has become obsolete. Apart from the dominant role which states will continue to play in environmental regimes, economic actors possess some potential which they can use to influence the distribution of values in developing countries in connection with regime policies. In the long term, developing countries can come more under pressure to reduce pollution or to make a more substantial contribution to conservation measures particularly in those regimes where industrialized countries have first taken major responsibility for the implementation of such measures. Improved effectiveness of domestic governance in developing countries could certainly contribute to reduce pressure on ecosystems. On the one hand, special responsibility for problem-solving that arises from a country's proportionally high contribution to a problem has predominantly been considered in the context among industrialized countries where the length of national coasts and riversides or their shares in trans-boundary pollution is reflected in the costs occurring for the implementation of environmental measures. On the other hand, such responsibilities were less important for the development of regime policies when they affected the behavior of developing countries. Empirical analysis reveals that the detection of distributional consequences of regimes on its own is difficult enough. The international community began to consider that socio-economic disparities between developing countries and industrialized countries must be taken into account when generating international environmental policies. Observations made for a number of regimes also revealed that this process has not yet been brought to an end.

Chapter 10
Non-State Actors and Participation in Regime Polities

One of the reasons given to justify obedience to international policies refers to those rights and rules established by a regime polity which govern participation by non-state actors. How did these rules historically emerge in international institutions? In what way have the political activities of non-state actors caused the creation of such procedures? How far do decision-making procedures of international regimes consider the participatory claims of an emerging global civil society? Do these procedures reflect the normative requirements of equality and reciprocity, openness, or discursiveness? Before we can deal with these questions, it should be noted that a great portion of the empirical data which have been collected for answering these questions are not included in the IRD. The design of the IRD was initially driven by the intention to explore whether a number of explanations for the formation or effectiveness of environmental regimes can be maintained, will be disproved, or must be changed when confronted with empirical findings that are based on a set of cases exceeding the small number of case studies normally used for the development or testing of theoretical approaches. The developers of this database are aware of the fact that many questions that will be researched in connection with this database can sufficiently be answered only by using additional information that is not included in the database itself.

The IRD does not contain detailed information about rules governing participation by non-state actors in regimes. Compared to the set of cases included in the IRD the collection of supplementary data about these rules covered a less comprehensive set of regimes. Information about participation rules has been collected with a structured questionnaire which was sent to more than 30 regime secretariats. Detailed information concerning formal and informal participation rules could be gathered for a total of 13 regimes of which a subset of eight regimes has also been coded for the IRD (see Table 10.1). Further on, some regime secretariats which avoided filling out this questionnaire submitted materials which inform on the topic. The supplementary data refer to participation rules which apply to non-state actors other than inter-governmental organizations. The preferential status existing for international organizations in environmental regimes results in the granting of rights which normally go beyond the scope of those rights existing for other non-state actors. Therefore, the recommendations for further expansion of these rights arising from this empirical study primarily aim at participation rules which apply for non-state actors other than intergovernmental organizations. In the following, three steps will be taken to answer the above-mentioned questions:

First, it will be explored whether regimes established rules for participation and which conditions non-state actors will have to fulfill for getting accepted as observers. It will be illustrated that a general trend towards the formalization of rules governing participation of non-state-actors could be observed until the end of the twentieth century. Second, it will be analyzed which participation rights have been established in regimes. It will also be explored whether non-state actors make use of the rights granted in international regimes and how decision-making and the management of regimes is determined by non-state actors in practice. Third, these rights and rules or forms of participation will be analyzed in terms of whether they fulfill the normative requirements that constitute transnational public spheres. It can be established that improvements took place in environmental regimes concerning the fulfillment of these normative requirements. But these improvements are less pronounced in decision-making bodies than in working groups or in bodies dealing with the programmatic activities of environmental regimes. In a sense, the creation of norms and rules that reflect the readjustment of the relationship between international society and emerging global civil society in international institutions progressed. However, this process must be further advanced so that environmental regimes will come closer to realizing those normative requirements which are applied to transnational systems of rule.

The Evolution of Participation Rights

Informal rules concerning non-state actor participation were applied in international negotiations early on – long before the League of Nations was established. For the nineteenth and early twentieth century, Steve Charnovitz (1997) highlights that participation of non-state actors was frequently accepted or even encouraged by state delegations when they negotiated new international conventions on the rules of war, intellectual property, health and labor issues or nature protection. Intergovernmental negotiations were, of course, determined by governments, but when "general multilateral conferences were held, NGOs invited themselves" (Charnovitz 1997, 212). The formalization of participation rules gained momentum in the twentieth century.[1] Non-state actors called for the creation of regimes and for the strengthening of international norms or participated in designing and implementing international programs for the management of trans-boundary problems that were launched under the authority of international institutions. Many of these actors focused their activities not only on agenda-setting, on mobilizing transnational concerns and on lobbying with political decision-makers. They also contributed their expertise during the formation or management of international institutions (Bäckstrand 2006). The evolution of scientific progress, the production of various kinds of technical or other standards in the global economy,

---

1   For studies which describe the influence of various types of non-state actors in environmental negotiations see Beisheim (2004) and Brühl (2003).

or improvements in health care and education or basic living conditions in the developing world could only be achieved by relying on the support of these actors (Barrett and Frank 1999; Chabbott 1999; Loya and Boli 1999; Schofer 1999). The constitution of the International Organization of Labor formally included in the 1919 Peace Treaties of Versailles specified the probably most far-reaching formal rules about non-state actor participation that emerged until then. According to article 389 of the constitution, the General Conference should "be composed of four Representatives of each of the Members, of whom two shall be Government Delegates and the two other shall be Delegates representing respectively the employers and the workpeople of each of the Members".[2] In the Covenant of the League of Nations only Article 25 directly dealt with non-state actors. It called for the support for the establishment of national Red Cross Organizations.

The formalization of participation rules took on a new quality with the creation of the 1945 Charter of the United Nations, where Article 71 offers the legal framework for non-state actor participation within the Economic and Social Council (ECOSOC). It provides for ECOSOC to "make suitable arrangements for consultation with non-governmental organizations which are concerned with matters within its competence".[3] Steve Charnovitz (1997, 253) argues that Article 71 "set a benchmark for other U.N. agencies". Only in the late 1960s or early 1970s serious efforts have been taken to make these provisions more concrete. Further progress has been made in this matter with ECOSOC-Resolutions 1296 of 23 May 1968 and E/1996/31 of 25 July 1996. Both resolutions take different degrees of involvement and competence displayed by non-state actors as criteria for distinguishing three categories of NGO-participation (general consultative status, special consultative status, Roster) in the work of the Economic and Social Council of the United Nations. However, the ECOSOC-Resolutions 1296 and E/1996/31 also reflect the claim of international society that non-state actors are subordinate and that states which are not members of the Council of ECOSOC will be given preferential status as observers. The ECOSOC-Resolution 1296 explicitly points out that "the arrangements for consultation should not be such as to accord to non-governmental organizations the same rights of participation as are acceded to States not members of the Council and to specialized agencies brought into relationship with the United Nations". It is also established in the document that the arrangements for consultation with non-state actors "should not be such as to overburden the Council or transform it from a body for coordination of policy and action, as contemplated in the Charter, into a general forum for discussion".[4]

---

2   See Part XII of the Peace Treaties of Versailles, Saint-Germain and Trianon; Part XII of the Peace Treaty of Neuilly of June 28, 1919, reprinted in Knipping, von Mangoldt and Rittberger (1996, 1456).

3   See Charter of the United Nations of June 26, 1945 (as amended on December 20, 1971) reprinted in Knipping, von Mangoldt and Rittberger (1995, 48).

4   See part II of ECOSOC-Resolution 1296 (XLIV) of May 23, 1968. Identical wordings are included in ECOSOC-Resolution E/1996/31 of July 25, 1996.

In many regimes formalization of participation rules did not gain momentum before the mid-1980s. Kal Raustiala (1997, 723) infers from more recently concluded international environmental treaties that they include rules that provide access of NGOs to policymaking processes, whereas "most of the pre-1985 multilateral treaties are silent on the subject of NGO access or grant only the most limited and easily rescinded access". While this finding is true until the mid-1990s, a number of international regimes that emerged more than two decades ago began to establish rules of procedure or guidelines that consider the topic of participation by non-state actors. The creation of such rules is a response to rising demands of various types of non-state actors to get access to international regimes.

To what extent have rules governing participation of non-state actors been established in those regimes for which supplementary data have been collected? In 11 of the 13 regimes represented in our subset formal participation rules have emerged until the end of the past decade. Decision-making in these regimes is based on more or less comprehensive rules of procedure governing, among other things, participation by non-state actors. For example, the rules that exist for the international whaling regime provide that "any international organization with offices in more than three countries may be represented at meetings of the Commission by an observer or observers, if such international organization has previously attended any meeting of the Commission, or if it submits its request in writing to the Commission 60 days prior to the start of the meeting and the Commission issues an invitation with respect to such request". Further on, these rules include that observers are "admitted to all meetings of the Commission and the Technical Committee, and to any meetings of subsidiary groups of the Commission and the Technical Committee, except the Commissioners-only meetings and the meetings of the Finance and Administration Committee" (IWC 2001, 2).[5] Some regimes developed specific guidelines concerning participation that were added to the rules of procedure. Regimes that are embedded in broader institutional settings like international organizations can rely on rules of procedure that have been developed for the organization as a whole. For example, the Executive Body of the LRTAP-regime agreed at its first session in 1983 to adopt the rules of procedure of ECE which include provisions for participation of accredited NGOs. These rules are applied fairly liberal by the Executive Body and its three main subsidiary bodies. Task forces, workshops and expert groups have no agreed rules but traditionally welcome NGO participation.

Roughly a third of the total set of cases included in the IRD is represented in this subset. The broad majority of regimes which has been coded for the IRD, but is not represented in this subset, also established rules that govern participation of non-state actors. For example, the regime that emerged with the 1966 International Convention for the Conservation of Atlantic Tunas now provides

---

5   The rules of procedure for the 1946 International Convention for the Regulation of Whaling conceive of a non-state actor as an "international organization" and distinguishes it from an "intergovernmental organization".

additional guidelines and criteria for granting an observer status that expand those rules established in the Convention or in its rules of procedure. International organizations, scientific organizations, companies and individual fishermen, or international fisheries commissions have been referred to in various articles of the Convention or in the provisions of the rules of procedure as possible observers or collaborators. The newly established guidelines explicitly provide that "all non-governmental organizations (NGOs) which support the objectives of ICCAT and a demonstrated interest in the species under the purview of ICCAT should be eligible to participate as an observer".[6]

These findings support the conclusion that environmental regimes are determined by a general trend towards the formalization of participation rules. But a number of examples also reveal that this development is not equally true for all environmental regimes or that it occurred with a temporal delay in individual cases. Eleven regimes of our subset experienced a formalization of rules. The regime on cooperation for the sustainable development of the Mekong River Basin is a special case. Only a number of informal rules exist that were not yet formalized by the end of 2002. Another case included in this subset reveals that there are still regimes where neither formal nor informal rules emerged concerning the participation of non-state actors. The prevention of water pollution in the Lake Constance is managed by an international commission whose work is very much determined by annual meetings of high-level officers from Austria, Germany and Switzerland as countries bordering on Lake Constance. During the more than 40 years of the commission's existence sub-national and national governments participating in these meetings succeeded in gearing the work of the commission to the conception of high-level officers and in avoiding that interest-groups would influence decision-making of the commission. This strategy led to the depoliticization of the commission's work. It involved that non-state actors were less interested in the work of the commission and influenced its work only indirectly via domestic channels which they used to put forward their attitudes toward water pollution in the Lake Constance. Accordingly, neither formal nor informal rules emerged in this regime until the end of the twentieth century.

All of the 12 regimes with formal or informal participation rules provide that non-state actors must be acknowledged or registered (e.g., by the secretariat or other regime bodies like the Conference of the Parties) or are subject to other accreditation procedures. Further on, these regimes established criteria on the basis of which non-state actors can be granted or denied access to, or participation in, meetings of the regime. Many regimes established criteria which refer to the transnational character or to the functional role of non-state actors in the issue-area. For example, rule 7 of the rules of procedure for the so-called Ramsar Convention of 1971 on Wetlands provides that "any body or agency, national or international,

---

6 See "Guidelines and Criteria for Granting Observer Status at ICCAT Meetings", adopted by the Commission at its 11th Special Meeting, Santiago de Compostela, November 16–23, 1998.

whether governmental or non-governmental, qualified in fields relating to the conservation and sustainable use of wetlands (...) may be represented at the meeting by observers, unless at least one third of the Parties present at the meeting object".[7]

While the functional role of non-state actors in the issue-area represents the most important criterion for granting the status as an observer, additional criteria are frequently used to govern participation. For example, the International Commission for the Protection Order agreed upon guidelines on the admission of non-state actors as observers that are more demanding than those formal and informal rules which govern participation in the regimes that are included in our subset. An international treaty concluded between Germany, Poland, the Czech Republic and the European Union in 1996 governs the environmental management of this river which flows into the Baltic Sea. The guidelines developed in 2002 provide that the activities of non-state actors must have a trans-boundary character or be more than just regionally oriented. They also prescribe that the work of non-state actors must refer to water protection, that these actors must possess expert knowledge concerning the issues dealt with by the Commission, and that they must acknowledge the goals and basic principles of the treaty.[8] Further on, these guidelines also demand that non-state actors were given a mandate for their activities from their own members or from the bodies used for internal decision-making. This requirement is not unusual for international institutions, even though it is not explicitly mentioned in the rules of procedure of many environmental regimes. While both ECOSOC-resolutions include such a requirement, their regulations do not apply to the rules of procedure of our subset of environmental regimes. The more recent resolution which has been adopted in 1996 further specified this requirement by establishing that non-state actors "shall have a representative structure and possess appropriate mechanisms of accountability to its members, who shall exercise effective control over its policies and actions through the exercise of voting rights or appropriate democratic and transparent decision-making".[9]

Though it would be desirable that international regimes would generally demand of non-state actors to submit to the requirement of a democratic mandate, the implementation of such a requirement in regimes may encounter resistance

---

7   See "Rules of Procedure for Meetings of the Conference of the Parties to the Convention on Wetlands of International Importance Especially as Waterfowl Habitat (Ramsar, Iran 1971) adopted by the 7th Meeting of the Conference of the Contracting Parties, San José, Costa Rica, 10–18 May 1999". A similar rule is also included in the rules of procedure for meetings of the Conference of the Parties to the Convention on Biological Diversity.

8   See "Grundsätze für die Zulassung von internationalen und nationalen Organisationen als Beobachter bei der Internationalen Kommission zum Schutz der Oder gegen Verunreinigung".

9   See part I of ECOSOC-Resolution E/1996/31 of July 25, 1996.

from countries with authoritarian political systems. The requirement itself may improve the consciousness of non-state actors that transnational democracy is based on the fulfillment of democratic procedures on various levels, which also include internal decision-making by these actors. While many regimes prescribe that non-state actors must acknowledge its basic goals and principles it could be argued that this requirement is only justified if the basic provisions of a regime coincide with the preferences of the transnational public on regime-related issues. Many of the basic goals and principles of environmental regimes certainly meet the demands of emerging global civil society. However, it can not be excluded that the basic character of a regime or some of its principles will contradict with the attitudes of transnational civil society. On the one hand, such a requirement could be used by states to keep critics away from arenas of global policymaking. On the other hand, it can protect functional-sectoral environmental orders against groups which intend to substantially weaken their basic goals sometimes.

The broad majority of our subset of regimes does not distinguish between different categories of participation (general consultative status, special consultative status, Roster) specified in the two ECOSOC-Resolutions of 1968 and 1996. Only within the rules governing participation in the LRTAP-regime such a distinction is made. On balance, this finding suggests that the observer status given to non-state actors provides participation rules that apply to all observers equally. Nevertheless, on the practical level of policymaking such a distinction is frequently made and also necessary when regime bodies rely on the special expertise of single non-state actors during regime management. The regime that emerged with the 1971 Ramsar Convention on Wetlands granted four non-state actors official status as international partners (Birdlife International, IUCN, WWF, Wetlands International) which may participate in big meetings as observers and in expert meetings as members, whereas all other non-state actors must demonstrate their role in the conservation and wise use of wetlands before they can participate as observers.

The regime on the conservation of wetlands can also be referred to as an example which illustrates that participation rules increasingly consider stakeholders on the local level. In 1999, the Conference of the Contracting Parties to the Convention on Wetlands adopted guidelines which intend to further improve involvement of local communities and indigenous people in the management of wetlands. These guidelines emerged from 23 commissioned case studies which revealed "that local and indigenous people's involvement can, if carried out within the full framework of the Convention, contribute significantly to maintaining or restoring the ecological integrity of wetlands, as well as contributing to community well-being and more equitable access to resources".[10] Four of the regimes included in our subset include regulations which provide that participation rights are transitory and not valid forever. They must be renewed from time to time. In some regimes

---

10 See Resolution VII.8 and its Annex adopted by the 7th Meeting of the Conference of the Contracting Parties to the Convention on Wetlands (Ramsar, Iran 1971), San José, Costa Rica, May 10–18, 1999.

it is also explicitly provided that non-adherence to the conditions for granting observer status can lead to suspend this status for an observing non-state actor.

The Existence and Exercise of Participation Rights

The prime hypothesis according to which the legitimacy of international regimes can only be improved if non-state actors can bring their potential to bear has some implications for the rights that must be provided by a regime polity. These rights can not be reduced to formal decision-making procedures. Rather, the regime polity has to guarantee that non-state actors can contribute to realizing a number of other grounds for legitimacy that are not directly connected with the demand for transnational democracy. Accordingly, the regime polity must provide participation rights which also guarantee that non-state actors can contribute to reduce various forms of uncertainties, to improve compliance by states and problem-solving or goal-attainment, or to achieve equitable outcomes. Under the circumstances, transnational democracy is also an intervening variable between the activities of non-state actors and the production of various outputs, outcomes and impacts. Only analysis of the concrete rights that govern participation in discourse and decision-making of regimes or which regulate participation in the programmatic activities for the achievement of various outcomes can inform about the extent to which participation is possible. The existence of such rights and their exercise by non-state actors in regimes will be explored in two steps. First, attention will be paid to rights which govern participation in decision-making bodies of regimes. Second, analysis will focus on rights governing participation in those programmatic activities of regimes which contribute to the production of outcomes and impacts. In the subset of regimes it will also be explored whether non-state actors exercise these rights. Moreover, it will be analyzed whether the exercise of participation rules by non-state actors impaired the achievement of authoritative decisions in regimes.

Rights Governing Participation in Decision-Making

Our subset of 13 regimes will be explored with regard to whether six specific rights have been implemented that govern participation by non-state actors in regimes: the granting of observer status, the right to access documents of decision-making bodies, the right to speak in meetings of decision-making bodies, the right to submit written proposals to state delegates within these bodies, the right to suggest own topics for the agenda, and suffrage in decision-making bodies. Most of these rights have already been specified in ECOSOC-Resolutions 1296 and E/1996/31. On the basis of a distinction between three categories of participation, the resolutions govern the granting of the status as an observer, of the right to propose items of special interest to non-state actors for the agenda, of the right to submit written statements, and of a right to present an oral statement to the Council or a sessional committee of the Council of ECOSOC.[11] If a regime polity acknowledges transnational societization as a matter of fact, it will grant the six rights described above which guarantee that deliberation on regime-related matters will not only reflect the political interests of international society but also the attitudes or interests of transnational stakeholders. In some ways, the above-described six rights are arranged in an order which implies an increase in the democratic quality of participation rules. In this order, the granting of the status as an observer is a minimum right without which other rights could not be exercised by non-state actors. In addition, the right to access documents is also prerequisite to exercise a number of other rights that govern participation in regime discourse. Table 10.1 includes the findings about participation rights of non-state actors in decision-making bodies of regimes.

In eight of the 11 regimes which provide formal participation rules the observer status applies to all decision-making bodies existing in a regime (see Table 10.1). Three regimes limit the observer status to selected decision-making bodies. Nevertheless, a closer look at single regimes which actually provide observer status to all decision-making bodies reveals that access to working groups which make no final decisions can be limited. For example, the North Atlantic Salmon Conservation Organization (NASCO) established conditions for non-government observers at NASCO-meetings which prescribe that the "observer status shall apply to all plenary sessions of the Council and Commissions, whether they be at the Annual Meeting or at inter-sessional meetings". However, these conditions exclude participation as observers in "meetings of NASCO's Working Groups or Committees".[12]

Information collected from regime secretariats reveals that non-state actors do in fact exercise their rights as observers. All regimes of our subset which provide participation rules gave non-state actors the right to access the documents of all decision-making bodies. In a few cases, full access can be constrained when documents are marked confidential. This right is exercised by non-state actors in each of these cases. The Mekong River Commission grants such a right on an informal basis, but non-state actors have limited access to documents. In general, access to documents about regime-related topics has improved during the past decade. Regime secretariats responded to the expansion of the Internet and used this newly emerging global communication medium to inform the transnational public more extensively. But for many non-state actors or individuals from developing countries access to the Internet is still constrained as a result of lacking electronic equipment and infrastructure or restrictions imposed by governments

---

11   See particularly Parts III–V of ECOSOC Resolution 1296 (XLIV).

12   See North Atlantic Salmon Conservation Organization: Conditions for Non-Government Observers at NASCO Meetings (mimeo).

Table 10.1 Participation Rights for Non-State Actors in Decision-Making Bodies of Regimes

	Regime included in IRD	Formal/ Informal Rules	Criteria Governing Access to or Participation in Regime/Various Categories of Participation	Observer Status/Access to Documents in all or Selected Decision–making Bodies	Right to Submit Written Proposals/Right to Suggest Topics for all or selected Decision–making Bodies	Right to Speak/Right to Vote in all or selected Decision–making Bodies
Stratospheric Ozone Regime	+	Formal	+/–	+/Full Access	–/–	+/–
Climate Change Regime	+	Formal	+/–	+/Full Access	+/–	+/–
Biodiversity Regime	+	Formal	+/–	+/Full Access	+/–	+/–
Regime for the International Regulation of Whaling	+	Formal	+/–	+/Full Access	–/–	–/–
IATTC–Regime (Inter–American Tropical Tuna Commission)	+	Formal	+/–	+/Full Access	+/–	+/–
Ramsar Regime (Convention on Wetlands)	+	Formal	+/–	+/Full Access	+/+	+/–
Regime on Long–Range Transboundary Air Pollution in Europe	+	Formal	+/+	+/Full Access	+/–	–/–
Baltic Sea Regime	+	Formal	+/–	+/Full Access	+/–	+/–
NASCO–Regime (North Atlantic Salmon Commission)	–	Formal	+/–	+/Full Access	+/+	+/–
Berne Convention on the Conservation of European Wildlife and Natural Habitats	–	Formal	+/–	+/Full Access	+/+	+/–
Convention on the Conservation of Migratory Species of Wild Animals	–	Formal	+/–	+/Full Access	+/–	+/–
International Commission for Water Protection of the Lake Constance	–	No rules	n.a.	n.a.	n.a.	n.a.
Mekong River Commission	–	Informal	+/–	–/Limited Access	–/–	–/–

+ (yes) – (no)

intending to prevent participation of their societies in transnational exchange of information.

A right to speak in all decision-making bodies has been established by nine regimes, but only in six of them this right takes a formal character. Such rights can be limited. For example, the formal rules established by the North Atlantic Salmon Conservation Organization prescribe that NGOs can make statements at the opening session of the Council and at the sessions defined as special sessions by the council. In addition, the formal rules adopted by the organization prescribe that "one joint five-minute statement may be made at the Opening Session of each Commission Meeting". Three regimes provide for a right to speak at meetings of selected decision-making bodies. In each of the regimes which provide a right to speak non-state actors make use of this right. In seven regimes of our subset, non-state actors have the right to submit written proposals to state delegates in meetings of all decision-making bodies. While in four of these seven cases this right has an informal character, it is applied by non-state actors in each of these cases. The LRTAP-secretariat communicated that non-state actors have the right to submit written proposals but do not often take advantage of this. The right to suggest own topics for the agenda of meetings is less developed in these regimes. Only two of them provide formal or informal rights to influence the agenda of all decision-making bodies, whereas the North Atlantic Salmon Conservation Organization (NASCO) provides such a right for selected meetings. Non-state actors have been consulted by the organization on topics for special sessions of the Council. A more comprehensive right that applies to all decision-making bodies has been formally established in the regime that emerged with the 1971 Ramsar Convention on Wetlands and exists on an informal basis in the regime created with the Berne Convention on the Conservation of European Wildlife and Natural Habitats. Both regimes have to rely on the expertise of non-state actors that are active in the field of nature conservation to realize the goals specified in the legal documents which constitute these institutions.

These findings illustrate that in one way or the other distinguished non-state actors can in fact influence the agenda of decision-making bodies in regimes when contributing to various regime functions. However, in the majority of the cases included in our subset such rules have not yet been formalized. The less non-state actors will be able to contribute to programmatic activities of regimes, the less will they be able to directly influence the agendas of regime bodies. This finding is not unusual. Rather, the domestic level is determined by similar differences with regard to the ability of single non-state actors to get access to or determine the agendas of governmental commissions or agencies. Finally, none of the regimes in our subset provided full or limited suffrage for non-state actors in decision-making bodies. As far as participation in decision-making bodies is concerned, only the minimum rights governing observer status and access to documents seem to be sufficiently realized by the regimes included in our subset. The majority of these regimes conceded a right to speak and a right to submit written proposals to non-state actors, but both rights often still had an informal character or were applied in a fairly restrictive way. In these regimes the right to set own topics on the agenda was poorly developed. Obviously, states have been receptive to the creation of participation rules inasmuch as these rights did not undermine the dominant role of states in regimes. Thus, it is not surprising that none of these regimes provided voting rights for non-state actors.

Rights Governing Participation in the Operation of Regimes

If non-state actors shall be able to participate in the implementation of regime policies, the regime polity has to provide rights which govern participation by these actors. Such rights include participation in scientific monitoring, in research on causes and effects of the problems managed by a regime, in compliance monitoring, in the verification of compliance, in the review of implementation, in reviewing the adequacy of commitments, or in information management. The majority of rights granted to non-state actors for participation in the programmatic activities of regimes are informal. This is partly caused by the fact that the rules of procedure or the guidelines that govern participation are often not very specific but leave room for interpretation. Further on, when programmatic activities have informal character or operate on an ad hoc basis participation rules are also informal. Information about the establishment of such rules is available for at least 11 regimes, and ten of them did in fact establish such rules. It should be noted that most of these regimes established only a subset rather than the full range of programmatic activities. Eight regimes provided rules which granted participation in the scientific monitoring of the causes and effects of environmental problems. In most of these regimes, non-state actors in fact contributed to monitoring. Rules about participation in research about the causes and effects of a problem existed for eight regimes likewise, and seven regimes have developed rules for both forms of participation. In nearly each of these regimes, non-state actors made use of these rights. The Antarctic Treaty system can be used to further illustrate the important role played by non-state actors for monitoring or scientific research. Admittedly, scientific research in this institutional setting remained very much determined by organizations on the national level. Nevertheless, the Scientific Committee on Antarctic Research (SCAR) which was established by the International Council of Scientific Unions (ICSU) "served as a central organ of the Treaty system capable of mobilizing considerable resources to achieve cooperative scientific objectives" on the international level (Herr 1996, 97) during the entire period since the Antarctic Treaty has been concluded in 1959. Further on, SCAR has been given a similar role in other legal documents that constitute the Antarctic Treaty system such as the 1964 Agreed Measures for the Conservation of Fauna and Flora, the 1972 Convention for the Conservation of Antarctic Seals, or the 1991 Protocol on Environmental Protection of the Antarctic Treaty system.[13] Obviously, the changing

---

13   A similar finding results from Lee A. Kimball's assessment about the role of NGOs in regimes which are associated or closely linked with the 1982 UN Convention on the Law

character of major other NGOs in the Antarctic Treaty system from agenda-setters to supporters and facilitators which helped to improve the effectiveness of the regime finally led to substantial revisions concerning participation rules. Richard Herr (1996, 106) illustrates that the rules of procedure of the Antarctic Treaty system which were revised in 1992 "gave all NGOs much more liberal access to the Consultative Meetings' agenda items so that, at the XVIII Consultative Meeting, held in Kyoto in 1994, NGOs were barred only from discussions on organizational arrangements at the opening and closing of Consultative Meetings and from debates on the proposed ATS secretariat".

Rules about participation by non-state actors in the verification of compliance existed in four regimes. Six regimes of our subset provided rules governing participation in the review of the adequacy of commitments specified by the rules of a regime. Five regimes provided rules for participation in both compliance monitoring and review of implementation and in four of these regimes non-state actors contributed to these regime functions. The impact of systems for implementation review (SIRs) on the behavior of regime members have been explored comprehensively by a collaborative project carried out under the leadership of David G. Victor and Eugene Skolnikoff. SIRs are conceived of as institutions which operate at the international level and "through which the parties share information, compare activities, review performance, handle noncompliance, and adjust commitments" (Victor, Raustiala and Skolnikoff 1998b, 3). SIRs consist of both formal and informal instruments through which implementation can be reviewed or compliance be assessed. Since many of the procedures which belong to these SIRs are informal or will be carried out ad hoc, many of the rules which govern participation by non-state actors in the operation of these review systems are also informal by nature.

The demand for reliable information arising in connection with the operation of SIRs can particularly be fulfilled by scientific and technical service organizations, international organizations, or those economic target groups which are affected by international regulation (Mitchell, Clark and Cash 2006). Thus, non-state actors which contribute to the functioning of SIRs have monitoring facilities and scientific or technical expertise at their disposal which allow to assess implementation or compliance reliably. In addition, such contributions to SIRs can normally only be made if these actors became engaged in this field on a fairly long-term basis. With regard to the effective implementation of wildlife regimes Peter H. Sand (2001, 45) illustrates the "crucial importance of independent monitoring schemes to ensure the scientific and political credibility of a regime, given the notorious unreliability

---

of the Sea. The author illustrates that many NGOs "are engaged in projects in the field: testing new approaches, ground-truthing more theoretical research and working hand-in-hand with local communities to improve knowledge and know-how". Further on, the author points out that these actors "organise international coalitions and workshops to promote improved ocean policies and practices, and to collaborate with IGOs in project and program implementation" (Kimball 1999, 391).

of governmental self-reporting". Against this background, the author points out that non-state actors contribute to two distinct varieties of monitoring mechanisms. Non-state actors like SCAR or the IUCN's World Conservation Monitoring Centre (WCMC) contribute to *resource monitoring* which assesses the conservation status of resources. Further on, empirical evidence for a long-term involvement of non-state actors in *compliance monitoring* has been provided already for the CITES-regime, where the TRAFFIC-network contributes to effectively implement and to control compliance with wildlife trade regulations.

Rules concerning the participation of non-state actors in information management existed in seven regimes. In one way or the other, non-state actors contributed to information management by disseminating information about the regime to a broader public or by producing and proliferating special reports. In at least five regimes of our subset single non-state actors operated as contractors under the regime and received payments for specific services which they provided. Among these cases are the regimes on the conservation of European wildlife and natural habitats (Berne Convention), on wetlands (Ramsar-Convention), on the conservation of biodiversity and on the conservation of migratory species of wild animals. These regimes deal with nature conservation or the management or sustainable use of natural resources. Finally, in two of these regimes single non-state actors contributed to fund single regime activities.

Does participation by non-state actors make the work of regime bodies less efficient? On balance, the empirical findings concerning the functioning of regime-wide bodies suggest the conclusion that participation by these actors does not constrain the effectiveness of decision-making in regimes. For example, data which reveal whether regime-wide bodies that are called for by the regime's constitutive agreements are in operation and produce authoritative decisions on a regular as-needed basis are available for 430 regime bodies (e.g., regime-wide bodies such like the Conference of the Parties, technical working groups, ad-hoc groups or scientific advisory boards) coded for IRD. These bodies belong to nearly 150 regime elements. The broad majority of 422 answers referring to the functioning of regime bodies in regime elements confirmed that these bodies which emerged from the constitutive legal agreements of regimes were in fact in operation. In 380 of the 422 cases where regime bodies were in operation these bodies also produced authoritative decisions on a regular as-needed basis. In general, these data support the conclusion that international regimes in fact produce decisions after they have been established. International environmental institutions do not only exist on paper, but political discourse or the work of different programmatic activities eventually cause regime members to arrive at decisions – apart from whether these decisions are far-reaching enough or how long the process of decision-making will finally be in single cases.

The above-described data also reveal that growing influence of non-state actors on political processes or on the management of environmental regimes did not cause serious problems for collective decision-making in regimes. The filters established by regimes to avoid paralysis of decision-making that could be caused by participation of non-state actors are normally not so permeable that the functioning of various regime bodies would be impaired. While the criteria which govern access to and participation in environmental regimes by non-state actors are obviously not too lax, the question arises on whether these participation rules are too restrictive at least in some cases that are part of our subset. Obviously, a number of wildlife regimes which have more comprehensive rules concerning participation by non-state actors in regime management benefit from non-state expertise in various ways. Non-state actors accept the division of labor that arises from the functional differences existing between various types of actors. The practical skills which they can offer determine whether they participate in the implementation of various regime functions and which contributions they will make in this regard or whether they will focus on advocacy and on lobbying for their political goals.

Participation Rights and Normative Requirements

It can be diagnosed that in principle the international society of states has acknowledged the demand for increased participation of non-state actors in regimes. However, there is a significant degree of uncertainty with regard to how far-reaching participation rights should be implemented in international institutions. While the emerging global civil society wrestles with international society in international institutions for the expansion of participation rights, constitutionalization on the level beyond the nation-state can be understood as an evolutionary process. It involves an experimental character where the feasibility of new participation rules is tried out step by step. Thus, the granting of participation rights which took on a new quality during the past two decades is an improvement, even though these rights may not be far-reaching enough in many cases. Regimes will learn from one another with respect to the practicality of participation rules. Our conception of transnational democracy builds on discourse as a means through which consensus can be reached between states or many different transnational interest-groups. A discursive conception of democracy involves that a number of normative requirements must be fulfilled before the outcomes of deliberation deserve obedience. In the following, attention will focus on whether participation rules meet the normative criteria of equality and reciprocity, openness, or discursiveness. This assessment will provide the basis for the suggestion that further improvements in discursive democracy on the level beyond the nation-state can be achieved if those rights will be extended which refer to participation by non-state actors in political debate or in the shaping of the agendas of decision-making bodies.

The empirical results of our previous analysis of participation rules suggest the conclusion that states have a great advantage over non-state actors concerning the contributions which they can make in political discourse in decision-making bodies. In working groups or bodies managing the various programmatic activities of regimes, discourse between non-state actors and states normally resembles much more the ideal of reciprocity or equality because only fulfillment of these requirements will guarantee the quality of deliberation on scientific or technical issues. This empirical finding coincides with the thesis developed by Peter Willetts (2000, 93) who argues that the "smaller the decision-making body, the lower its public profile, the more technical the subject matter, and the more experienced the NGO representatives, the more likely it becomes that the NGOs can take a full part in the discussions and exercise significant influence". Since scientists or other types of experts can make a functional contribution to assessments, they are privileged vis-à-vis other non-state actors with regard to access to distinct regime bodies. Such preferential treatment can be justified as long as the work of these actors does not lead to the rule of a new transnational expertocracy sealing itself off from political discourse in the broader transnational public sphere. The common practice of disseminating assessment reports to the broader public involves that these assessments will be examined by other experts or be confronted with preferences existing on the transnational public sphere. Many examples could be given which can illustrate that assessment reports which emerged from the work of regime bodies are reviewed among experts, will be influenced by knowledge other than scientific, or be discussed in the transnational public sphere. For example, the assessments made by various bodies which are dealing with scientific, technical, or socio-economic issues in the ozone regime or the climate change regime arise from a process where a broad number of research institutes contribute their expertise or review the drafts produced by these bodies (Litfin 1994; Parson 2003). Of course, the contributions which experts can make to problem-solving sets them apart from other non-state actors in the context of the work carried out in programmatic activities. But other non-state actors can act as a corrective which contributes to harmonize the views shared by experts with the policies preferred by the transnational public. Thus, fears according to which a transnational expertocracy would impose its will on world society are for the most part unfounded in the field of international environmental governance.

Scientific experts must not inevitably be more influential than other non-state actors during the various stages of the policy cycle. A study produced by Elizabeth Corell (1999) on the negotiations of the 1994 United Nations Convention to Combat Desertification (UNCCD) revealed that the International Panel of Experts on Desertification (IPED) was less influential in this process than one would expect from its role as the formally appointed scientific advisory body. To the contrary, NGOs had a higher influence on these negotiations because of the coherence of NGO-networks, the supportive environment in the negotiations, or the expertise which they could provide about the impact of desertification on the local level. Further on, the bottom-up-approach in future implementation discussed as a central legal element of this convention legitimized NGO-participation in negotiations from the beginning. The author concludes "that the concept of 'expert', as usually employed in many studies of expertise, should be expanded to include not only scientists, but also other actors who possess relevant knowledge"

(Corell 1999, 216). Against this background, it can be concluded that many of those actors which are not participating in the programmatic activities of regimes can nevertheless contribute to epistemic processes by reviewing, criticizing or supplementing these activities. Programmatic activities not inevitably cause an empowerment of experts, even though one has to admit that the management of highly complex issues is susceptible to the evolution of an oligarchy of experts. If necessary, assessments or policy proposals produced by regime experts will meet with opposition by other segments of transnational civil society and become part of transnational public debate.

The absence of voting rights for non-state actors in our subset illustrates that states hold nearly exclusive authority in decision-making. However, in many regimes decisions are mainly achieved by consensus. Decision-making by consensus can avoid that a majority would outvote a minority of states. Majority decisions are in many cases inappropriate for regimes since they can not guarantee the willingness of defeated states to obey. They bear the danger that particularly defeated states will lack capacities to implement policies agreed upon by the majority. Therefore, it can be argued that in the many cases where decision-making in regimes is not determined by majority decisions the granting of voting rights for non-state actors is currently less urgent than the granting of rights which guarantee participation by non-state actors in discourse. Against this background, further rights must be implemented in order to reduce the asymmetries which characterize the relationship between non-state actors and states in discourse of decision-making bodies. Priority should be given to the expansion of the right to speak or the right to influence agendas in decision-making bodies. It is hardly conceivable that equality between states and non-state actors can be achieved within the next few decades. However, the restrictions existing with regard to the provision of the right to speak in decision-making bodies prevent that discourse will be more determined by reciprocity and be less dominated by states. Thus, it seems advisable to change participation rules gradually and to try out whether the expansion of rights for non-state actors to speak in decision-making bodies or to influence agenda-setting are practicable. The optimistic assessment given by Peter Willetts (2000, 206) that the change in the legal norms on participation which occurred in the UN-system during the past decade is "revolutionary because it implies an equality of status between governments and NGOs" or that these actors "have become a third category of subjects in international law, alongside states and intergovernmental organizations" should not blind us too much. While some progress has been made concerning the rules that govern participation by non-state actors, from a normative perspective this process has not yet been brought to an end.

The findings further suggest that equal status has been granted to all actors in nearly each of the regimes explored in this subset. Multinational firms or industrial associations have similar participation rights than non-profit environmental non-state actors or scientific or technical service organizations. Political processes are often characterized by under-representation of non-state actors from developing countries. Unequal participation by actors from the Northern or Southern hemisphere is not caused by participation rules but by differences in resources which favor NGOs from the developed world vis-à-vis groups from developing countries.

Decision-making bodies or programmatic activities became more open for non-state actors in the course of time. Some environmental regimes have also begun to become more aware of the role which local stakeholders can play in the management of resources or consider the right of self-determination of local groups. In some respects, for advocacy organizations participation rules primarily serve to guarantee their role as institutionalized opposition in global governance systems. The data hardly allow to assess how far the work of various programmatic activities or working groups that exist in regimes have been determined by discursiveness. The only judgment that can be made starts from the finding that regimes provide many bodies which lay the epistemic foundations on the basis of which decision-makers can develop policies which deal with individual problems. The dialogue among scientific and technical experts must be based on discursiveness since the quality of consensual knowledge depends on the exchange among scientists and technical experts. Data included in the IRD revealed the relevance of co-action between science-based and bureaucratic communities. In 164 regime elements bureaucratic communities with a common policy enterprise which were located in government agencies, international organizations, non-governmental organizations were identified by coding case study experts as expert groups which were present and active during the political process. For 149 regime elements science-based communities located in national or international research institutes or research departments were identified as present and active. For 145 regime elements, both types of expert groups became active simultaneously during the political process. These findings suggest that deliberation in regimes has in fact been strongly determined by a mix between scientific rationality and policy-expertise, but the breakthrough of social reason has also been constrained by differences in state-interests. Less ambiguous data are available for the role of principled beliefs communities and of expert groups who are based on legal beliefs. Both types of expert groups have been mentioned for more than 100 regime elements, but inter-coder reliability is constrained by the fact that disagreement exists between some coders concerning the relevance which can be ascribed to these communities in some regimes. While for science-based communities only seven critical answers exist where the second coder avoided the coding of this answer, much less convergence exists concerning the data which were provided for the latter two types of expert groups. The comments provided by coders reveal that the four types of expert groups are overlapping. For example, expert groups which are based on legal beliefs can be considered as part of bureaucratic communities likewise. Members of principled beliefs communities who share a set of normative and principled beliefs which provide a value-based rationale for social action can also belong to bureaucratic or science-based communities.

Conclusion

International regimes have attracted the attention of transnational societal actors, but these regimes are still far from providing all those rights for participation which these actors demand. The six rights which have been explored with regard to whether they are provided by a regime polity and practiced by non-state actors are minimum rights. Regime discourse is still dominated by international society. At the moment, possible fears that increased participation by non-state actors in regime discourse could degenerate them into debating societies are for the most part unfounded. It may be too early to fulfill each of these rights immediately. A discursive conception of transnational democracy builds on rights which guarantee that transnational civil society will be given the opportunity to involve the international society of states into an exchange of arguments on the rightness and appropriateness of international policies. While some efforts must be made by states to embody these rights in regimes, non-state actors are faced with the challenge to adjust their internal decision-making procedures in such a way that their representatives in international governance systems will become accountable to their members or to all those other individuals which they represent. Participation in a permanent discourse of the transnational public spheres causes new requirements for the organizational structure of a non-governmental organization. This applies particularly to nonprofit-oriented non-state actors which act on behalf of the public interest. Long-lasting participation in regime discourse demands that many of these organizations will have to regularize their representation in meetings of regime bodies. The interaction of many problems in world politics involves that discourse on regime-related issues can no longer be fenced within clearly de-bordered issue-areas. Thus, non-state actors are confronted with the task to adjust their work to partly contradictory requirements. The more complex and intense political processes will become in a regime, the more efforts will have to be made by a non-state actor so that it can exercise participation rights comprehensively and contribute to regime discourse in a way that fulfills the expectations of its members.

Chapter 11

Conclusion

The notion of "governance without government" has been used by students of international politics for illustrating that no government exists beyond the level of the nation-state which could exert authority on international regimes. This formula directed our view on global juridification as a mainly non-hierarchical form of governance that avoids the creation of a world state, but it has drawn off attention from the character of international regimes as systems of political rule in world society. The absence of a global state or state-like forms of international government beyond the territory of the European Union does not speak against the relevance of legitimacy for the evolution of social order in world society. Power and self-interest alone can hardly guarantee that social order can emerge or be maintained beyond the national level. Those governance systems which are based on the power of a state or state-coalitions can be kept alive in the long-term only if other grounds exist that convince world society of their usefulness and of the consideration of demands for equity and participation. The mode of legitimacy does not disqualify the use of power. But it implies that power can only play a productive role in world politics if its exercise can be legitimized by other grounds. Legitimacy does not conflict with the assumption that outcomes and impacts arising from global governance systems will have to fulfill the self-interest of various actors in world society. However, it goes beyond a purely rationalist account of global governance. It considers that international institutions developed to systems of political rule where the willingness to follow policies is not only contingent on utilitarian considerations. National constituencies and transnational communities became increasingly aware of the evolution of new spheres of authority which affect their well-being. The protest against international institutions that was frequently raised in the transnational public sphere can not purely be understood as a rebellion against lacking opportunities for participation of emerging global civil society. These actors express the concerns of emerging global civil society and demand for regime policies to become more effective. The standards that are used for assessing the outcomes and impacts of policies and decision-making procedures on the domestic level of liberal states apply in a similar way to governance systems that exist beyond the nation-state.

Under the circumstances, the willingness to obey to the policies of international regimes is contingent on the fulfillment of a number of grounds for legitimacy. These grounds take normative requirements just as much into account as the utility which these institutions are expected to provide in world society. These standards are not only applied by various actors of emerging global civil society, but also members of international society themselves make an assessment on whether various grounds justify to follow the policies of individual institutions. It would not lead us very far if only a binary distinction between effective and ineffective outcomes and impacts would be applied in the development of reasonable judgments on whether obedience to institutions is justified. Since various factors that relate to the character of an institution, that consider the activities of non-state actors, or which exist outside the domain of a regime can contribute that outcomes and impacts of global governance can change for the better, the more interesting task seems to be to identify those conditions which can in fact accomplish that the state of the world can be improved (whether these improvements refer to changes in the constellation of interests among social actors or to improvements in the state of the environment).

Much attention has been paid by students of global governance to the design of international institutions as a crucial factor for the production of effective policies. By and large, coding case study experts ascribed an important causal role to international regimes for improvements in the state of the social and natural worlds. One has to admit that differences exist among the regimes which were explored with regard to the level of improvements that could be achieved. However, empirical findings also revealed that factors located outside of the domain of a regime mainly accounted for lacking improvements in certain issue-areas. This finding indicates that external factors can operate as constraints for problem-solving by an institution. In addition, this finding raises the question on whether the impact of an institution on regime consequences remains relatively stable over time or whether the stimulus that came from an institution at a particular moment caused a development involving that the production of future changes will occur more independently from a regime. The interactions that exist between institutions in world politics also remind us that the ability of a regime to solve a problem can depend on the character of policies that are implemented by neighboring international institutions or by private governance systems.

The results that emerged from the quantitative study of regimes suggest that consensual knowledge about cause-effect-relationships or about the use of policy options could be improved by the work of institutional mechanisms in regimes. Compliance mechanisms guaranteed that the deepening of cooperation did not negatively impact compliance behavior. They provided and assessed information about the state of national implementation or gave assistance to countries which had difficulties to implement international policies effectively. Mechanisms for financial or technology transfer facilitated implementation and compliance in developing countries or countries with economies in transition. They reduced possible asymmetries of the distributional consequences of environmental regimes. These changes in the social world were a precondition for improvements in the state of the environment in many regimes. In some respects, the provision of regimes with institutional mechanisms is a factor that impacts the fulfillment of various grounds for legitimacy. Thus, states have a special responsibility for providing international institutions with those mechanisms which are necessary that goals can be achieved or problems be solved. The causes for the non-attainment of goals or for ineffective problem-solving in world politics should not always be ascribed to the failure of institutions alone. States can be responsible for this failure when they refrain from providing international governance systems with those functional tasks and competencies which are needed for effective problem-solving.

The findings of this book clearly contradict realist claims that international institutions would, if at all, be a negligible factor for the creation of outcomes and impacts in world politics. It would certainly be unhelpful to continue the debate between neo-realists and institutionalists predominantly by the exchange of theoretical arguments, but realists are confronted with the challenge to provide us with empirical findings which can support their argument. Since knowledge about factors that account for the failure or demise of institutions is less advanced, further studies could improve our understanding in this matter. This would also contribute to the clarification on whether the demise or failure of international institutions mainly emerges as a result of the distribution of power among states in the international system, whether ineffective policies or the demise of institutions are mainly caused by institutional failure, or whether external factors that are located beyond the domain of a regime or of power distribution are responsible.

The internationalization of governance raises the question whether an adequate mix can be found between tasks and competencies that are primarily carried out internationally or by the nation-state. There is no need to carry out each regime function on the international level. The use of the subsidiarity principle can possibly reduce fears that the nation-state could suffer a further loss of its function as a collectivity which integrates territorially delimited social groups. In practice, the implementation of the principle of subsidiarity raises a number of tricky issues which will be one of the many topics that will become relevant in the general debate about the future constitution of the world polity. Growing depth of international cooperation that will become necessary from social, economic or cultural de-nationalization could involve that the transfer of authority to the level beyond the nation-state will be continued and that systems of rule that emerge beyond the level of the nation-state could become more government-like in particular cases.

Non-state actors are forces which can influence the evolution of social order in world politics in many respects. Empirical evidence has been found which supports the plausibility of the prime hypothesis. Obviously, regimes are determined by a similar development that can be observed on the level of the modern nation-state where a de-hierarchization in state-society-relationships became necessary for managing complex problems effectively and for securing political consensus in societies about the character of these policies. Activist non-state actors whose behavior is determined by contestation or civil disobedience for pressurizing states to strengthen environmental policies are not the only actors that are relevant for the production of outcomes and impacts. Service-oriented actors participated in the management of programmatic activities for the creation of consensual knowledge, for the management of compliance, or for capacity-building. Private firms have begun to realize the need to strengthen their own involvement in environmental regimes. Though many private firms were negatively affected from the policies of international regimes, they accepted in many cases their responsibility for the

conservation of natural resources or for the protection of other environmental goods
– whether this behavior has been caused by political pressure of environmentally
like-minded coalitions, by growing insight of firms into these necessities, or by
the expectations of environmentally like-minded firms to benefit from the demand
for environmentally-sound technologies. In some respects, regimes turned out as
meeting places where transnational discourse between various types of non-state
actors could take place on a regular basis.

While the political pressure of activist environmental groups has frequently
contributed to the implementation of more far-reaching environmental policies,
the real political divide existed not just between international society or non-state
actors. Within both camps actors could be found which supported or were opposed
to the development of environmental policies. The debate about the feasibility
of policies, about the availability of technological options for, or the economic
costs of problem-solving took place in programmatic activities of regimes and in
the broader transnational public. To this end, informal and formal participation
rights emerged within regimes. The gradual evolution of procedural rules for the
participation of non-state actors in regimes makes it possible to test the practicability
of these rules. From a normative perspective, the strengthening and formalization
of participation rules for non-state actors seems to be inevitable in many regimes.

The heterogeneity of interests in emerging global civil society or different
levels of expertise among non-state actors should also cause us to avoid over-
optimism concerning the potential of non-state actors. Non-state actors can be
only one among several factors which account for the legitimacy of international
regimes. Neither institutional mechanisms nor the important role of the state can
be ignored as factors that contribute to the production of outcomes and impacts.
States remained active and creative in regimes. They perceived regimes as
instruments which they can use for producing those outcomes and impacts which
they could not achieve by independent state action. The broad conception of the
term of "non-state actors" made it possible to consider various types of actors or
the functions carried out by them in global politics. The relationship between the
international society of states and emerging global civil society would be wrongly
understood as a zero-sum-game where the increase in the influence of one type
of actor involves a similar decrease in the influence of another type of actor. The
findings of this study suggest the conclusion that a partial increase of the influence
of non-state actors in regimes did not severely weaken the role of the nation-state
as an actor that still dominates political processes in environmental regimes in
many respects. The state realized that it will have to rely on non-state expertise for
improving the quality of, and for securing support for its own policies in issue-areas
that are affected by globalization. The diversification of actors in world politics
must not inevitably constrain state influence, but the discursive power of non-state
actors has partly been used by states as an additional resource for the attainment of
political goals in world politics. Theorists of globalization reputed the nation-state
to be dead or diagnosed that it would be in mortal agony. The nation-state is still
alive, but it had to adjust to the fundamental changes in world politics.

References

Abbott, K. W. and Snidal, D. (1998), 'Why States Act Through Formal International Organizations', *Journal of Conflict Resolution* 42:1, 3–32.

Adger, N., Paavola, J., Huq, S. and Mace, M. J. (eds) (2006), Fairness in Adaptation to Climate Change (Cambridge, MA: The MIT Press).

Adler, E. and Haas, P. M. (1992), 'Conclusion: Epistemic Communities, World Order, and the Creation of a Reflective Research Program', *International Organization* 46:1, 367–90.

Albin, C. (2001), *Justice and Fairness in International Negotiation* (Cambridge: Cambridge University Press).

Alkoby, A. (2006) 'Globalising a Green Civil Society: In Search of Conceptual Clarity', in Winter, G. (ed.), 106–46.

Andresen, S. (2002), 'The International Whaling Commission (IWC): More Failure Than Success?', in Miles, E. L., Underdal, A., Andresen, S., Wettestad, J., Skjaerseth, J. B. and Carlin, E. M. (eds), 379–29.

Andresen, S., Skodvin, T., Underdal, A. and Wettestad, J. (2000), *Science in International Environmental Regimes: Between Integrity and Involvement* (Manchester: Manchester University Press).

Archibugi, D. and Held, D. (1995), *Cosmopolitan Democracy: An Agenda for a New World Order* (Oxford: Polity Press).

Auer, M. and Nilenders, E. (2001), 'Verifying Environmental Cleanup: Lessons from the Baltic Sea Joint Comprehensive Action Programme', *Environment and Planning C: Governmental Policy* 19:6, 881–901.

Axelrod, R. and Keohane, R. O. (1986), 'Achieving Cooperation Under Anarchy: Strategies and Institutions', in Oye, K. A. (ed.), 226–54.

Bäckstrand, K. (2006) 'Democratizing Global Environmental Governance? Stakeholder Democracy after the World Summit for Sustainable Development', *European Journal of International Relations* 12:4, 467–98.

Bailey, J. L. (2008), 'Arrested Development: The Fight to End Commercial Whaling as a Case of Failed Norm Change', *European Journal of International Relations* 14:2, 289–318.

Barandat, J. (ed.) (1997), *Wasser – Konfrontation oder Kooperation: Ökologische Aspekte von Sicherheit am Beispiel eines weltweit begehrten Rohstoffs* (Baden-Baden: Nomos).

Barrett, D. and Frank, D. J. (1999), 'Population Control for National Development: From World Discourse to National Policies', in Boli, J. and Thomas, G. (eds), 198–221.

Beetham, D. (1991), *The Legitimation of Power* (Atlantic Highlands, NJ: Palgrave Macmillan).

206 The Legitimacy of International Regimes

Beisheim, M. (2004), Fit für Global Governance? Transnationale
Interessengruppenaktivitäten als Demokratisierungspotential – am Beispiel
Klimapolitik (Opladen: Leske + Budrich).

Beisheim, M., Dreher, S., Walter, G., Zangl, B. and Zürn, M. (1999), Im Zeitalter der
Globalisierung? Thesen und Daten zur gesellschaftlichen Denationalisierung
(Baden-Baden: Nomos).

Beisheim, M. and Zürn, M. (1999), 'Transnationale Nicht-
Regierungsorganisationen: Eine Antwort auf die Globalisierung?', in Klein,
A., Legrand, H. J. and Leif, T. (eds), 306–19.

Beitz, C. R. (1979), Political Theory and International Relations (Princeton, NJ:
Princeton University Press).

Benedick, R. E. (1991), Ozone Diplomacy: New Directions in Safeguarding the
Planet (Cambridge, MA: Harvard University Press).

Benhabib, S. (1995), 'Modelle des öffentlichen Raums. Hannah Arendt, die
liberale Tradition und Jürgen Habermas', in Benhabib, S., Selbst im Kontext:
Kommunikative Ethik im Spannungsfeld von Feminismus, Kommunitarismus
und Postmoderne (Frankfurt/M.: Suhrkamp), 96–130.

Benhabib, S. (1997), 'Die gefährdete Öffentlichkeit', Transit 13, 26–41.

Bergesen, H. O. and Parmann, G. (eds) (1997), Green Globe Yearbook 1997
(Oxford: Oxford University Press).

Bergesen, H. O., Parmann, G. and Thommessen, O. B. (eds) (1995), Green Globe
Yearbook 1995 (Oxford: Oxford University Press).

Bergesen, H. O., Parmann, G. and Thommessen, O. B. (eds) (1998), Yearbook
of International Co-operation on Environment and Development 1998/99
(London: Earthscan).

Bernauer, T. (1995), 'The Effect of International Environmental Institutions: How
We Might Learn More', International Organization 49:2, 351–77.

Bernauer, T. (1996), 'Protecting the River Rhine Against Chloride Pollution',
in Keohane, R. O. and Levy, M. A. (eds), Institutions for Environmental Aid
(Cambridge, MA: The MIT Press), 201–32.

Betsill, M. M. and Corell, E. (eds) (2007), NGO Diplomacy: The Influence of
Nongovernmental Organizations in International Environmental Negotiations
(Cambridge, MA: The MIT Press).

Beyerlin, U. (2000), Umweltvölkerrecht (München: C. H. Beck).

Beyme, K. v. (1997), Der Gesetzgeber. Der Bundestag als Entscheidungszentrum
(Opladen: Westdeutscher Verlag).

Biermann, F. (1998), Weltumweltpolitik zwischen Nord und Süd: Die neue
Verhandlungsmacht der Entwicklungsländer (Baden-Baden: Nomos).

Bodansky, D. (1999), 'The Legitimacy of International Governance: A Coming
Challenge For International Environmental Law?', American Journal of
International Law 93:3, 596–624.

Bohman, J. (1998), 'The Globalization of the Public Sphere', Philosophy & Social
Criticism 24:3, 199–216.

Bohman, J. (1999), 'International Regimes and Democratic Governance. Political Equality and Influence in Global Institutions', *International Affairs* 75:3, 499–513.

Boli, J. and Thomas, G. (eds), *Constructing World Culture: International Non-Governmental Organizations Since 1875* (Stanford: Stanford University Press), 100–126.

Bonacci, O. (2000), 'The Role of International Co-operation in a More Efficient, Sustainable Development of Water Resources Management in the Danube Basin', *European Water Management* 3:2, 26–34.

Bowman, M. J. (1995), 'The Ramsar Convention Comes of Age', *Netherlands International Law Review* XLII, 1–52.

Brecher, M. and Harvey, F. (eds) (2002), *Realism and Institutionalism in International Studies* (Ann Arbor: The University of Michigan Press).

Breitmeier, H. (1996), *Wie entstehen globale Umweltregime? Der Konfliktaustrag zum Schutz der Ozonschicht und des globalen Klimas* (Opladen: Leske + Budrich).

Breitmeier, H. (1997), 'International Organizations and the Creation of Environmental Regimes', in Young, O. R. (ed.), 87–114.

Breitmeier, H. (2004) 'International Governance and the Democratic Process: Consequences on Domestic and Transnational Levels', in Underdal, A. and Young, O. R. (eds), 281–306.

Breitmeier, H. (2006), 'Institutions, Knowledge and Change: Findings from the Quantitative Study of Environmental Regimes', in Winter, G. (ed.), 430–52.

Breitmeier, H., Levy, M. A., Young, O. R. and Zürn, M. (1996a) *International Regimes Database (IRD): Data Protocol* (Laxenburg: IIASA Working Paper WP-96-154).

Breitmeier, H., Levy, M. A., Young, O. R. and Zürn, M. (1996b): *The International Regimes Database as a Tool for the Study of International Cooperation* (Laxenburg: IIASA Working Paper WP-96-160).

Breitmeier, H. and Rittberger, V. (2000), 'Environmental NGOs in an Emerging Global Civil Society', in Chasek, P. S. (ed.), 130–63.

Breitmeier, H., Young, O. R. and Zürn, M. (2006), *Analyzing International Environmental Regimes: From Case Study to Database* (Cambridge, MA: The MIT Press).

Brooks, L. A. and VanDeveer, S. D. (eds) (1997), *Saving the Seas: Values, Scientists, and International Governance* (Maryland: University of Maryland Sea Grant Publications).

Brühl, T. (2003), *Nichtregierungsorganisationen als Akteure internationaler Umweltverhandlungen* (Campus Verlag: Frankfurt/Main).

Brühl, T., Debiel, T., Hamm, B., Hummel, H. and Martens, J. (eds) (2001), *Die Privatisierung der Weltpolitik: Entstaatlichung und Kommerzialisierung im Globalisierungsprozess* (Bonn: J. H. W. Dietz).

Brunkhorst, H. and Niesen, P. (eds) (1999), *Das Recht der Republik* (Frankfurt/M.: Suhrkamp).

Bull, H. (1977), *The Anarchical Society: A Study of Order in World Politics* (Houndmills: Columbia University Press).

Buzan, B. (1993), 'From International System to International Society: Structural Realism and Regime Theory Meet the English School', *International Organization* 47:3, 327–52.

Calließ, J. and Moltmann, B. (eds) (1995), *Die Zukunft der Außenpolitik: Deutsche Interessen in den internationalen Beziehungen* (Loccum: Evangelische Akademie).

Carlsnaes, W., Risse, T. and Simmons, B. A. (eds) (2002), *Handbook of International Relations* (London: Sage).

Cavender-Bares, J., Jäger, J. and Ell, R. (2001), 'Developing a Precautionary Approach: Global Environmental Risk Management in Germany', in The Social Learning Group (ed.) (2001a), 61–91.

Chabbott, C. (1999), 'Development INGOs', in Boli, J. and Thomas, G. (eds), 222–48.

Charnovitz, S. (1997), 'Two Centuries of Participation: NGOs and International Governance', *Michigan Journal of International Law* 18:2, 183–286.

Chasek, P. S. (ed.) (2000), *The Global Environment in the Twenty-Fist Century: Prospects for International Cooperation* (Tokyo: United Nations University Press).

Chatfield, C. (1997), 'Intergovernmental and Nongovernmental Associations to 1945', in Smith, J., Chatfield, C. and Pagnucco, R. (eds), 19–41.

Chayes, A. and Handler Chayes, A. (1993), 'On Compliance', *International Organization* 47:2, 175–205.

Chayes, A. and Handler Chayes, A. (1995), *The New Sovereignty: Compliance with International Regulatory Agreements* (Cambridge, MA: Harvard University Press).

Chayes, A., Handler Chayes, A., Mitchell, R. B. (1998), 'Managing Compliance: A Comparative Perspective', in Weiss, E. B. and Jacobsen, H. K. (eds), 39–62.

Clark, W. C., van Eijndhoven, J. and Dickson, N. M. et al. (2001), 'Option Assessment in the Management of Global Environmental Risks', in The Social Learning Group (ed.) (2001b), 49–85.

Clark, W. C., Mitchell, R. B. and Cash, D. W. (2006), 'Evaluating the Influence of Global Environmental Assessments', in Mitchell, R. B., Clark, W. C., Cash, D. W. and Dickson, N. M. (eds), 1–28.

Claude, I. L. Jr. (1966), 'Collective Legitimization as a Political Function of the United Nations', *International Organization* 20:3, 367–379.

Cochran, M. (1999), *Normative Theory in International Relations: A Pragmatic Approach* (Cambridge: Cambridge University Press).

Cohen, J. L. and Arato, A. (1992), *Civil Society and Political Theory* (Cambridge, MA: The MIT Press).

Corell, E. (1999), 'Non-State Actor Influence in the Negotiations of the Convention to Combat Desertification', *International Negotiations* 4, 197-223.

Crane, A., McWilliams, A., Matten, D., Moon, J. and Siegel, D. S. (2008), *The Oxford Handbook of Corporate Social Responsibility* (Oxford: Oxford University Press).

Cutler, A. C. (2001), 'Critical Reflections on the Westphalian Assumption of International Law and Organization: A Crisis of Legitimacy', *Review of International Studies* 27:2, 133–50.

Dahl, R. A. (1994), 'A Democratic Dilemma: System Effectiveness versus Citizen Participation', *Political Science Quarterly* 109:1, 23–34.

Dai, X. (2007), *International Institutions and National Policies* (Cambridge: Cambridge University Press).

Downs, G. W., Rocke, D. M. and Barsoom, P. N. (1996), 'Is the Good News About Compliance Good News About Cooperation?', *International Organization* 50:3, 379–406.

Economic and Social Council (ECOSOC) (1968), ECOSOC Resolution 1296 (XLIV) – Arrangements for Consultation with Non-Governmental Organizations, New York.

Economic and Social Council (ECOSOC) (1996), ECOSOC Resolution E/1996/31 – Consultative Relationship between the United Nations and Non-governmental Organizations, New York.

Efinger, M., Mayer, P. and Schwarzer, G. (1993), 'Integrating and Contextualizing Hypotheses: Alternative Paths to Better Explanations of Regime Formation', in Rittberger, V. (ed.), 252–81.

Efinger, M., Rittberger, V. and Zürn, M. (1988), *Internationale Regime in den Ost-West-Beziehungen: Ein Beitrag zur Erforschung der friedlichen Behandlung internationaler Konflikte* (Frankfurt/M: Haag und Herchen).

van Eijndhoven, J., Clark, W. C. and Jäger, J. (2001), 'The Long-Term Development of Global Environmental Risk Management: Conclusions and Implications for the Future', in The Social Learning Group (ed.) (2001a), 181–97.

Evans, T. and Wilson, P. (1992), 'Regime Theory and the English School of International Relations: A Comparison', *Millenium* 21:3, 329–51.

Fearon, J. D. (1996), 'Causes and Counterfactuals in Social Science: Exploring an Analogy between Cellular Automata and Historical Processes', in Tetlock, P. and Belkin, A. (eds), 39–67.

Finlayson, J. A. and Zacher, M. W. (1988), *Managing International Markets: Developing Countries and the Commodity Trade Regime* (New York: Columbia University Press).

Finnemore, M. (1999), 'Rules of War and Wars of Rules: The International Red Cross and the Restraint of State Violence', in Boli, J. and Thomas, G. M. (eds), 149–65.

Franck, T. M. (1995), *Fairness in International Law and Institutions* (Oxford: Clarendon Press).

Franck, T. M. (1992), 'The Emerging Right to Democratic Governance', *American Journal of International Law* 86:1, 46–91.

Frank, D. J., Hironaka, A., Meyer, J. W., Schofer, E. and Tuma, N. B. (1999), 'The Rationalization and Organization of Nature in World Culture', in Boli, J. and Thomas, G. M. (eds), 81–99.

Frank, D. J., Hironaka, A., Schofer, E. (2000), 'The Nation-State and the Natural Environment Over the Twentieth Century', *American Sociological Review* 65:1, 96–116.

Gale, F. P. (1998), *The Tropical Timber Trade Regime* (Basingstoke: Palgrave Macmillan).

Goldstein, J. and Keohane, R. O. (1993), *Ideas and Foreign Policy: Beliefs, Institutions and Political Change* (Ithaca: Cornell University Press).

Goodin, R. E. (1996), 'Institutions and Their Design', in Goodin, R. E. (ed.), *The Theory of Institutional Design* (Cambridge: Cambridge University Press), 1–53.

Gordenker, L. and Weiss, T. G. (1996), 'Pluralizing Global Governance: Analytical Approaches and Dimensions', in Weiss, T. G. and Gordenker, L. (eds), *NGOs, the UN & Global Governance* (Boulder: Lynne Rienner), 17–47.

Greene, O. (1998), 'Implementation Review and the Baltic Sea Regime', in Victor, D. G., Raustiala, K. and Skolnikoff, E. B. (eds), 177–220.

Grote, J. R. and Gbikpi, B. (eds) (2002), *Participatory Governance. Political and Societal Implications* (Opladen: Leske + Budrich).

Haas, E. B. (1990), *When Knowledge is Power: Three Models of Change in International Organizations* (Berkeley, CA: University of California Press).

Haas, P. M. (1990), *Saving the Mediterranean: The Politics of International Environmental Cooperation* (New York: Columbia University Press).

Haas, P. M. (1992), 'Introduction: Epistemic Communities and International Policy Coordination', *International Organization* 46:1, 1–35.

Haas, P. M. (1993), 'Protecting the Baltic and the North Seas', in Haas, P. M., Keohane, R. O. and Levy, M. A. (eds), 133–81.

Haas, P. M., Keohane, R. O. and Levy, M. A. (eds) (1993), *Institutions for the Earth: Sources of Effective International Environmental Protection* (Cambridge, MA: The MIT Press).

Habermas, J. (1992), *Faktizität und Geltung. Beiträge zur Diskurstheorie des Rechts und des demokratischen Rechtsstaats* (Frankfurt/M.: Suhrkamp).

Habermas, J. (1998), *Die postnationale Konstellation: Politische Essays* (Frankfurt/M.: Suhrkamp).

Hall, J. A. (1996), *International Orders* (Cambridge: Polity Press).

Hasenclever, A., Mayer, P. and Rittberger, V. (1997), *Theories of International Regimes* (Cambridge: Cambridge University Press).

Haufler, V. (1997) 'Dancing with the Devil: International Business and Environmental Regimes', in Brooks, L. A. and VanDeveer, S. D. (eds), 309–80.

Held, D. (1995), *Democracy and the Global Order: From the Modern State to Cosmopolitan Governance* (Cambridge: Polity Press).

Held, D. (1997), 'Democracy and Globalization', *Global Governance* 3:3, 251–67.

Held, D. and McGrew, A. (1998), 'The End of the Old Order? Globalization and the Prospects for World Order', *Review of International Studies* 24:3, 219–43.

Helm, C. and Sprinz, D. (1999), *Measuring the Effectiveness of International Environmental Regimes* (Potsdam: Postdam Institute for Climate Impact Research, PIK Report No. 52).

Helsinki Commission (Helcom) (1999), *Annual Report 1999 of Helcom Programme Implementation Task Force* (PITF) (Helsinki: Helcom).

Herr, R. A. (1996), 'The Changing Roles of Non-Governmmental Organisations in the Antarctic Treaty System', in Stokke, O. S. and Vidas, D. (eds), 91–110.

Hisschemöller, M. and Gupta, J. (1999), 'Problem-solving through International Environmental Agreements: The Issue of Regime Effectiveness', *International Political Science Review* 20: 2, 151–74.

Höffe, O. (1997), 'Für und wider eine Weltrepublik', *Internationale Zeitschrift für Philosophie* 6:2, 218–33.

Holsti, K. J. (1991), *Peace and War: Armed Conflicts and International Order 1648-1989* (Cambridge: Cambridge University Press).

Honneland, G. (1998), 'Compliance in the Fishery Protection Zone Around Svalbard', *Ocean Development & International Law* 29:4, 339–60.

Huckel, C., Rieth, L., Zimmer, M. (2007), 'Die Effektivität von Public-Private Partnerships', in Hasenclever, A., Wolf, K.-D. and Zürn, M. (eds) *Macht und Ohnmacht internationaler Institutionen* (Campus Verlag: Frankfurt/Main), 115–44.

Hurd, I. (1999), 'Legitimacy and Authority in International Politics', *International Organization* 53:2, 379–408.

Hurrell, A. (1993), 'International Society and the Study of Regimes: A Reflective Approach', in Rittberger, V. (ed.), 49–72.

Hurrell, A. and Kingsbury, B. (eds) (1992), *The International Politics of the Environment: Actors, Interests, and Institutions* (Oxford: Clarendon Press).

Inter-American Tropical Tuna Commission (IATTC) (1999), *Annual Report of the Inter-American Tropical Tuna Commission 1997* (La Jolla, CA: IATTC).

International Commission for the Conservation of Atlantic Tuna (ICCAT) (1998), 'Guidelines and Criteria for Granting Observer Status at ICCAT Meetings', adopted by the Commission at its 11th Special Meeting, Santiago de Compostela – November 16 to 23, 1998 (mimeo).

International Commission for the Protection of the Rhine (ICPR) (2001), 'Rules of Procedure and Financial Regulations of the ICPR as last amended by the 67th Plenary Assembly of the ICPR', July 3, 2001 in Luxemburg (mimeo).

International Whaling Commission (IWC) (2001), 'Rules of Procedure and Financial Regulations' (as amended by the Commission at the 53rd Annual Meeting, July 2001) (mimeo).

Internationale Kommission zum Schutz der Oder gegen Verunreinigung (2002), 'Grundsätze für die Zulassung von internationalen und nationalen Organisationen als Beobachter bei der Internationalen Kommission zum Schutz der Oder gegen Verunreinigung', (mimeo).

Jacobsen, H. K. (1984), *Networks of Interdependence: International Organizations and the Global Political System*, 2nd Edition (New York: McGraw-Hill).

Jacobson, H. K. and Weiss, E. W. (1998), 'A Framework for Analysis', in Weiss, E. B. and Jacobsen, H. K. (eds), 1–18.

Joseph, J. (1994), 'The Tuna-Dolphin Controversy in the Eastern Pacific Ocean: Biological, Economic and Political Impacts', *Ocean Development and International Law* 25:1, 1–30.

Joyner, C. C. (1996), 'The Effectiveness of CRAMRA', in Stokke, O. S. and Vidas, D. (eds), 152–73.

Joyner, C. C. (1998), *Governing the Frozen Commons: The Antarctic Regime and Environmental Protection* (Columbia, CA: University of South California Press).

Joyner, C. C. (1999), 'The Concept of the Common Heritage of Mankind in International Law', *Emory International Law Review* 13:2, 615–28.

Joyner, C. C. (2000), 'The Legal Status and Effect of Antarctic Recommended Measures', in Shelton, D. (ed.), *Commitment and Compliance: The Role of Non-Binding Norms in the International Legal System* (Oxford: Oxford University Press), 163–95.

Kaelble, H., Kirsch, M. and Schmidt-Gernig, A. (eds) (2002), *Transnationale Öffentlichkeiten und Identitäten im 20. Jahrhundert* (Frankfurt/M.: Campus).

Kaldor, M. (2000), '"Civilising" Globalisation? The Implications of the "Battle in Seattle"', *Millenium* 29:1, 105–14.

Kaspar, M. (1999), *Erfolgsfaktoren regionaler Umweltprogramme in Mittel- und Osteuropa* (Wien: Wuv Universitätsverlag).

Keck, M. E. and Sikkink, K. (1998), *Activists Beyond Borders. Advocacy Networks in International Politics* (Ithaca: Cornell University Press).

Kenis, P. and Schneider, V. (1991), 'Policy Networks and Policy Analysis: Scrutinizing a New Analytical Toolbox', in Marin, B. and Mayntz, R. (eds), 25–59.

Keohane, R. O. (1984), *After Hegemony: Cooperation and Discord in the World Political Economy* (Princeton, NJ: Princeton University Press).

Keohane, R. O. (2001), 'Governance in a Partially Globalized World. Presidential Address, American Political Science Association, 2000', *The American Political Science Review* 95:1, 1–13.

Keohane, R. O. and Martin, L. (1995), 'The Promise of Institutionalist Theory', *International Security* 20:1, 39–51.

Keohane, R. O., Moravcsik, A. and Slaughter, A. M. (2000), 'Legalized Dispute Resolution: Interstate and Transnational', *International Organization* 54:3, 457–88.

Keohane, R. O. and Nye, J. S. Jr. (1998), 'Power and Interdependence in the Information Age', *Foreign Affairs* 77:5, 81–94.

Kimball, L. A. (1999), 'Major Challenges of Ocean Governance: The Role of NGOs', in Vidas, D. and Ostreng, W. (eds), 389–405.

Kiss, A. C. (1996), 'Compliance with International and European Environmental Obligations', in *Hague Yearbook of International Law* (9), 45–54.

Klein, A., Legrand, H. J. and Leif, T. (eds) (1999), *Neue soziale Bewegungen: Impulse, Bilanzen und Perspektiven* (Opladen: Westdeutscher Verlag).

Knipping, F., von Mangoldt, H. and Rittberger, V. (eds) (1995), *Das System der Vereinten Nationen und seine Vorläufer Band I/1: Vereinte Nationen* (Bern: C. H. Beck).

Knipping, F., von Mangoldt, H. and Rittberger, V. (eds) (1996), *Das System der Vereinten Nationen und seine Vorläufer Band II: 19. Jahrhundert und Völkerbundszeit*, (Bern: C. H. Beck).

Kohl, A. (2002), 'Organisierte Kriminalität als NGO? – Die italienische Mafia', in Frantz, C. and Zimmer, A. (eds), *Zivilgesellschaft international: Alte und neue NGOs* (Opladen: Leske + Budrich), 329–46.

Koremenos, B., Lipson, C., Snidal, D. (2001a), 'The Rational Design of International Institutions', *International Organization* 55:4, 761–99

Koremenos, B., Lipson, C., Snidal, D. (2001b), 'Rational Design: Looking Back to Move Forward', *International Organization* 55:4, 1051–82.

Krasner, S. D. (1983), 'Structural Causes and Regime Consequences: Regimes as Intervening Variables', in Krasner, S. D. (ed.), 1–21.

Krasner, S. D. (ed.) (1983), *International Regimes* (Ithaca: Cornell University Press).

Kriesberg, L. (1997), 'Social Movements and Global Transformation', in Smith, J., Chatfield, C., Pagnucco, R. (eds), 3–18.

Krueger, J. (1996), *Regulating Transboundary Movements of Hazardous Wastes: The Basel Convention and the Effectiveness of the Prior Informed Consent (PIC) Procedure* (Laxenburg: IIASA Working Paper WP-96-113).

Levy, M. A. (1993), 'European Acid Rain: The Power of Tote-Board Diplomacy', in Haas, P. M., Keohane, R. O. and Levy, M. A. (eds), 75–132.

Levy, M. A. (1995), 'International Cooperation to Combat Acid Rain', in Bergesen, H. O., Parmann, G. and Thommessen, O. B. (eds), 59–68.

Levy, M. A., Keohane, R. O. and Haas, P. M. (1993), 'Improving the Effectiveness of International Environmental Institutions', in Haas, P. M., Keohane, R. O. and Levy, M. A. (eds), 397–426.

Levy, M. A., Young, O. R. and Zürn, M. (1995), 'The Study of International Regimes', *European Journal of International Relations* 1:3, 267–330.

Linnerooth-Bayer, J. and Murcott, S. (1996), 'The Danube River Basin: International Cooperation or Sustainable Development', *Natural Resources Journal* 36:3, 521–47.

List, M. and Rittberger, V. (1992), 'Regime Theory and International Environmental Management', in Hurrell, A. and Kingsbury, B. (eds), 85–109.

Litfin, K. T. (1994), *Ozone Discourses: Science and Politics in International Environmental Cooperation* (New York: Columbia University Press).

Loya, T. A. and Boli, J. (1999), 'Standardization in the World Polity: Technical Rationality Over Power', in Boli, J. and Thomas, G. M. (eds), 169–97.

Luterbacher, U. and Sprinz, D. (eds) (2001), *International Relations and Global Climate Change* (Cambridge, MA: The MIT Press).

Mansbridge, J. J. (1990), 'The Rise and Fall of Self-Interest in the Explanation of Political Life', in Mansbridge, J. J. (ed.), 3–22.

Mansbridge, J. J. (ed.) (1990), *Beyond Self-Interest* (Chicago: University of Chicago Press).

Marauhn, T. (1996), 'Towards a Procedural Law of Compliance Control in International Environmental Relations', *Zeitschrift für ausländisches öffentliches Recht und Völkerrecht* 56, 696–731.

March, J. G. and Olsen, J. P. (1989), *Rediscovering Institutions: The Organizational Basis of Politics* (New York: Free Press).

March, J. G. and Olsen, J. P. (1998), 'The Institutional Dynamics of International Political Orders', *International Organization* 52:4, 943–69.

Marin, B. and Mayntz, R. (eds) (1991), *Policy Networks: Empirical Evidence and Theoretical Considerations* (Frankfurt/M.: Campus).

Martin, L. L. and Simmons, B. A. (1998), 'Theories and Empirical Studies of International Institutions', *International Organization* 52:3, 729–57.

Mathews, J. T. (1997), 'Power Shift', *Foreign Affairs* 76:1, 50–66.

Mayer, P., Rittberger, V. and Zürn, M. (1993), 'Regime Theory: State of the Art and Perspectives', in Rittberger, V. (ed.), 391–430.

McAdam, D., McCarthy, J. D. and Zald, M. N. (1996), 'Introduction: Opportunities, Mobilizing Structures, and Framing Processes – Toward a Synthetic, Comparative Perspective on Social Movements', in McAdam, D., McCarthy, J. D. and Zald, M. N. (eds), 1–20.

McCarthy, J. D., Smith, J. and Zald, M. N. (1996), 'Accessing Public, Media, Electoral, and Governmental Agendas', in McAdam, D., McCarthy, J. D. and Zald, M. N. (eds), 291–311.

McAdam, D., McCarthy, J. D. and Zald, M. N. (eds) (1996), *Comparative Perspectives on Social Movements: Political Opportunities, Mobilizing Structures, and Cultural Framings* (Cambridge: Cambridge University Press).

Mearsheimer, J. J. (1995), 'The False Promise of International Institutions', *International Security* 19:1, 5–49.

Meinke, B. (2002), *Multi-Regime-Regulierung. Wechselwirkungen zwischen globalen und regionalen Umweltregimen* (Wiesbaden: Deutscher Universitätsverlag).

Miles, E. L., Underdal, A., Andresen, S., Wettestad, J., Skjaerseth, J. B. and Carlin, E. M. (eds) (2002), *Environmental Regime Effectiveness: Confronting Theory with Evidence* (Cambridge, MA: The MIT Press).

Mitchell, R. B. (1994), *Intentional Oil Pollution at Sea: Environmental Policy and Treaty Compliance* (Cambridge, MA: The MIT Press).

Mitchell, R. B. (2001), 'Which Environmental Treaties Work Best? Toward a Theory of Relative Effectiveness', Paper Prepared for the UCLA Political Study of International Law Speaker Series, 26 October 2001 (mimeo).

Mitchell, R. B., Clark, W. C. and Cash, D. W. (2006), 'Information and Influence', in Mitchell, R. B., Clark, W. C., Cash, D. W. and Dickson, N. M. (eds), 307–38.

Mitchell, R. B., Clark, W. C., Cash, D. W. and Dickson, N. M. (eds) (2006) *Global Environmental Assessments: Information and Influence* (Cambridge, MA: The MIT Press).

Neidhardt, F. (ed.) (1994), *Öffentlichkeit, öffentliche Meinung, soziale Bewegungen* [Special Issue 34 of the Kölner Zeitschrift für Soziologie und Sozialpsychologie] (Opladen: Westdeutscher Verlag).

Neidhardt, F. (1994), 'Öffentlichkeit, öffentliche Meinung, soziale Bewegungen', in Neidhardt, F. (ed.), 7–41.

North Atlantic Salmon Conservation Organization: Conditions for Non-Government Observers at NASCO Meetings (mimeo).

Oberndörfer, D. (1989), *Schutz der tropischen Regenwälder durch Entschuldung* (München: C. H. Beck).

Oberthür, S. (1998), 'The International Convention for the Regulation of Whaling: From Over-Exploitation to Total Prohibition', in Bergesen, H. O., Parmann, G. and Thommessen, O. B. (eds), 29–37.

Oberthür, S. (2001), 'Linkages Between the Montreal and Kyoto Protocols. Enhancing Synergies between Protecting the Ozone Layer and the Global Climate', *International Environmental Agreements: Politics, Law and Economics* 1:3, 357–77.

Oberthür, S., Buck, M. and Müller, S. (2002), *Participation of Non-Governmental Organisations in International Environmental Co-operation. Legal Basis and Practical Experience* (Berlin: Schmidt Verlag).

Oberthür, S. and Gehring, T. (eds) (2006): *Institutional Interaction in Global Environmental Governance: Synergy and Conflict among International and EU Policies* (Cambridge, MA: The MIT Press).

Oberthür, S. and Ott, H. (2000), *Das Kyoto-Protokoll: Internationale Klimapolitik für das 21. Jahrhundert* (Opladen: Leske + Budrich).

O'Brien, R., Goetz, A. M., Scholte, J. A. and Williams, M. (2000), *Contesting Global Governance: Multilateral Economic Institutions and Global Social Movements* (Cambridge: Cambridge University Press).

Olson, M. (1965), *The Logic of Collective Action: Public Goods and the Theory of Groups* (Cambridge, MA: Harvard University Press).

OSPAR Commission (1998), Convention for the Protection of the Marine Environment of the North-East Atlantic (mimeo).

Ostrom, E. (1990), *Governing the Commons: The Evolution of Institutions for Collective Action* (Cambridge: Cambridge University Press).

Oye, K. A. (ed.) (1986), *Cooperation under Anarchy* (Princeton, NJ: Princeton University Press).

Parson, E. A. (2003), *Protecting the Ozone Layer. Science and Strategy* (Oxford: Oxford University Press).

Parson, E. A. and Greene, O. (1995), 'The Complex Chemistry of the International Ozone Agreements', *Environment* 37:2, 16–20 and 34–43.

Paul, J. A. (2001), 'Der Weg zum Global Compact: Zur Annäherung von UNO und multinationalen Unternehmen', in Brühl, T., Debiel, T., Hamm, B., Hummel, H. and Martens, J. (eds), 104–29.

Peters, B. (1994), 'Der Sinn von Öffentlichkeit', in Neidhardt, F. (ed.), 42–76.

Peterson, C. L. and Bayliff, W. H. (1985), *Inter-American Tropical Tuna Commission. Special Report No. 5. Organization, Functions, and Achievements of the Inter-American Tropical Tuna Commission* (La Jolla, CA: IATTC).

Peterson, M. J. (1988), *Managing the Frozen South: The Creation and Evolution of the Antarctic Treaty System* (Berkeley, CA: University of California Press).

Peterson, M. J. (1997), 'International Organizations and the Implementation of Environmental Regimes', in Young, O. R. (ed.), 115–51.

Ramsar Convention (1999a), Report of the Secretary General, Ramsar COP7 DOC.5, 7th Meeting of the Conference of the Contracting Parties to the Convention on Wetlands (Ramsar, Iran, 1971), San José, Costa Rica, May 10–18, 1999.

Ramsar Convention (1999b), Rules of Procedure for Meetings of the Conference of the Parties to the Convention on Wetlands of International Importance Especially as Waterfowl Habitat (Ramsar, Iran 1971) adopted by the 7th Meeting of the Conference of the Contracting Parties, San José, Costa Rica, May 10–18, 1999.

Ramsar Convention (1999c), Guidelines for Establishing and Strengthening Local Communities and Indigenous People's Participation in the Management of Wetlands, Resolution VII.8 and its Annex adopted by the 7th Meeting of the Conference of the Contracting Parties to the Convention on Wetlands (Ramsar, Iran 1971), San José, Costa Rica, May 10–18, 1999.

Raustiala, K. (1997), 'States, NGOs, and International Environmental Institutions', *International Studies Quarterly* 41:4, 719–40.

Raustiala, K. (2001), 'Nonstate Actors in the Global Climate Regime', in Luterbacher, U. and Sprinz, D. (eds), 95–117.

Raustiala, K. and Victor, D. G. (1998), 'Conclusions', in Victor, D. G., Raustiala, K. and Skolnikoff, E. B. (eds), 659–707.

Rawls, J. (1979), *Eine Theorie der Gerechtigkeit* (Frankfurt/M.: Suhrkamp).

Reinalda, B. (2000), 'The International Women's Movement as a Private Political Actor Between Accomodation and Change', in Ronit, K. and Schneider, V., 165–86.

Reinicke, W. H. (1998), *Global Public Policy. Governing without Government?* (Washington, DC: Brooking Institution Press).

Risse, T. (2000), 'Let's Argue: Communicative Action in World Politics', *International Organization* 54:1, 1–39.

Risse, T., Ropp, S. C. and Sikkink, K. (eds) (1999), *The Power of Human Rights. International Norms and Domestic Change* (Cambridge: Cambridge University Press).

Rittberger, V. (ed.) (1990), *International Regimes in East-West Politics* (London: Pinter).

Rittberger, V. (ed.) (1993), *Regime Theory and International Relations* (Oxford: Clarendon Press).

Rittberger, V. (ed.) (2001), *Global Governance and the United Nations System* (Tokyo: United Nations University Press).

Rittberger, V. and Nettesheim, M. (eds) (2008), *Authority in the Global Political Economy* (New York: Palgrave Macmillan).

Rittberger, V. and Zangl, B. (2003), *Internationale Organisationen – Politik und Geschichte: Europäische und weltweite internationale Zusammenschlüsse*, 3rd edition (Opladen: Leske + Budrich).

Rittberger, V. and Zürn, M. (1990), 'Towards Regulated Anarchy in East-West Relations: Causes and Consequences of East-West Regimes', in Rittberger, V. (ed.), 9–63.

Röhrich, W. (ed.) (1967), *Macht und Ohnmacht des Politischen* (Cologne: Kiepenheuer & Witsch).

Ronit, K. and Schneider, V. (eds) (2000), *Private Organizations in Global Politics* (London: Routledge).

Rosenau, J. N. (1997), *Along the Domestic-Foreign Frontier: Exploring Governance in a Turbulent World* (Cambridge: Cambridge University Press).

Rucht, D. (1999), 'Gesellschaft als Projekt – Projekte in der Gesellschaft', in Klein, A., Legrand, H. J. and Leif, T., 15–27.

Ruggie, J. G. (1995), 'The False Promise of Realism', *International Security* 20:1, 62–70.

Sand, P. H. (1997), 'Commodity or Taboo? International Regulation of Trade in Endangered Species', in Bergesen, H. O. and Parmann, G. (eds) (1997), 19–36.

Sand, P. H. (2001), 'A Century of Green Lessons: The Contribution of Nature Conservation Regimes to Global Governance', *International Environmental Agreements: Politics, Law and Economics* 1:1, 33–72.

Sands, P. (1995), *Principles of International Environmental Law I: Frameworks, Standards and Implementation* (Manchester: Manchester University Press).

Sands, P., Tarasofsky, R. and Weiss, M. (eds) (1997), *Principles of International Environmental Law II: Documents in International Environmental Law* (Manchester: Manchester University Press).

Scharpf, F. W. (1998), *Interdependence and Democratic Legitimation* (Cologne: Working Paper 98/2, Max Planck Institut für Gesellschaftsforschung).

Scharpf, F. W. (1999), *Regieren in Europa: effektiv und demokratisch?* (Frankfurt/ M.: Campus).

Scharpf, F. W. (2000), *Interaktionsformen: Akteurszentrierter Institutionalismus in der Politikforschung* (Opladen: Leske + Budrich).

Schmitz, H. P. (2002), 'Nicht-staatliche Akteure und Weltöffentlichkeit: Menschenrechte in der zweiten Hälfte des 20. Jahrhunderts', in Kaelble, H., Kirsch, M. and Schmidt-Gernig, A. (eds), 423–41.

Schofer, E. (1999), 'Science Associations in the International Sphere, 1875–1990: The Rationalization of Science and the Scientization of Society', in Boli, J. and Thomas, G. (eds), 249–66.

Shapiro, M. (1993), 'The Globalization of Law', *Indiana Journal of Global Legal Studies* 1:1, 34–67.

Shaw, M. (2000), *Theory of the Global State. Globality as an Unfinished Revolution* (Cambridge: Cambridge University Press).

Singer, D. J. (1990), 'Variables, Indicators, and Data: The Measurement Problem in Macropolitical Research', in Singer, D. J. (ed.), 3–28.

Singer, D. J. (1990), *Models, Methods, and Progress in World Politics: A Peace Research Odyssey* (Boulder: Westview Press).

Skjaerseth, J. B. (2000), *North Sea Cooperation: Linking International and Domestic Pollution Control* (Manchester: Manchester University Press).

Slaughter, A. M. (1995), 'International Law in a World of Liberal States', *European Journal of International Law* 6:4, 503–38.

Smith, J., Chatfield, C. and Pagnucco, R. (eds) (1997), *Transnational Social Movements and Global Politics: Solidarity beyond the State* (Syracuse: Syracuse University Press).

Smith, J., Pagnucco, R. and Chatfield, C. (1997), 'Social Movements and World Politics', in Smith, J., Chatfield, C. and Pagnucco, R. (eds), 59–77.

Snidal, D. (2002), 'Rational Choice and International Relations', in Carlsnaes, W., Risse, T. and Simmons, B. A. (eds), 73–94.

The Social Learning Group (ed.) (2001a), *Learning to Manage Global Environmental Risks – Volume I: A Comparative History of Social Responses to Climate Change, Ozone Depletion, and Acid Rain* (Cambridge, MA: The MIT Press).

The Social Learning Group (ed.) (2001b), *Learning to Manage Global Environmental Risks – Volume II: A Functional Analysis of Social Responses to Climate Change, Ozone Depletion, and Acid Rain* (Cambridge, MA: The MIT Press).

Sternberger, D. (1967), 'Max Webers Lehre von der Legitimität', in Röhrich, W. (ed.), 110–26.

Stokke, O. S. (1996), 'The Effectiveness of CCAMLR', in Stokke, O. S. and Vidas, D. (eds), 120–51.

Stokke, O. S. (1998a), 'Nuclear Dumping in Arctic Seas: Russian Implementation of the London Convention', in Victor, D. G., Raustiala, K. and Skolnikoff, E. B. (eds), 457–517.

Stokke, O. S. (1998b), 'Beyond Dumping? The Effectiveness of the London Convention', in Bergesen, H. O., Parmann, G. and Thommessen, O. B. (eds), 39–49.

Stokke, O. S., Anderson, L. G. and Mitrovitskaya, N. (1999), 'The Barents Sea Fisheries Regime', in Young, O. R. (ed.), 91–154.

Stokke, O. S. and Vidas, D. (eds) (1996), *Governing the Antarctic: The Effectiveness and Legitimacy of the Antarctic Treaty System* (Cambridge: Cambridge University Press).

Strange, S. (1983), 'Cave! Hic Dragones: A Critique of Regime Analysis', in Krasner, S. D. (ed.), 337–68.

Streck, C. (2006), 'Financial Instruments and Cooperation in Implementing International Agreements for the Global Environment', in Winter, G. (ed.), 493–516.

Take, I. (2002), *NGOs im Wandel: Von der Graswurzel auf das diplomatische Parkett* (Wiesbaden: Westdeutscher Verlag).

Tallberg, J. (2002), 'Paths to Compliance: Enforcement, Management and the European Union', *International Organization* 56:3, 609–43.

Tandon, Y. (2001), 'Global Governance and Justice', in Rittberger, V. (ed.), 203–31.

Tarrow, S. (1998), *Power in Movement: Social Movements and Contentious Politics*, 2nd edition (Cambridge: Cambridge University Press).

Tetlock, P. and Belkin, A. (eds) (1996), *Counterfactual Thoughts Experiments in World Politics: Logical, Methodological, and Psychological Perspectives* (Princeton, NJ: Princeton University Press).

Teubner, G. (1999), 'Polykorporatismus: Der Staat als "Netzwerk" öffentlicher und privater Kollektivakteure', in Brunkhorst, H. and Niesen, P. (eds), 346–71.

Underdal, A. (1998), 'Explaining Compliance and Defection', *European Journal of International Relations* 4:1, 5–30.

Underdal, A. (2002a), 'One Question, Two Answers', in Miles, E. L., Underdal, A., Andresen, S., Wettestad, J., Skjaerseth, J. B. and Carlin, E. M. (eds), 3–45.

Underdal, A. (2002b), 'Conclusions: Patterns of Regime Effectiveness', in Miles, E. L., Underdal, A., Andresen, S., Wettestad, J., Skjaerseth, J. B. and Carlin, E. M. (eds), 433–65.

Underdal, A. and Young, O. R. (eds) (2004), *Regime Consequences: Methodological Challenges and Research Strategies* (Dordrecht: Kluwer Academic Publishers).

United Nations A/55/L.2, United Nations Millenium Declaration, 8th Plenary Meeting, 8 September 2000 (mimeo).

United Nations A/CONF.199/L.6/Rev.2, The Johannesburg Declaration on Sustainable Development, World Summit on Sustainable Development, Johannesburg, South Africa, August 26–September 4, 2002 (mimeo).

United Nations (Economic Commission for Europe) (1996), 1979 Convention on Long-Range Transboundary Air Pollution and its Protocols, New York and Geneva (mimeo).

United Nations Environment Programme (UNEP) (1996), *Register of International Treaties and Other Agreements in the Field of the Environment* (Nairobi: UNEP).

United Nations General Assembly A/AC.237.9, Report of the Intergovernmental Negotiating Committee for a Framework Convention on Climate Change on the Work of its Second Session, Held at Geneva June 19–28, 1991.

United States Fish & Wildlife Service (2003), African Elephant Conservation Fund, Information Sheet (mimeo).

Valiante, M., Muldoon, P. and Botts, L. (1997), 'Ecosystem Governance: Lessons from the Great Lakes', in Young, O. R. (ed.), 197–225.

Vanderheiden, S. (2008), *Atmospheric Justice: A Political Theory of Climate Change* (Oxford: Oxford University Press).

Victor, D. G. (1998), 'The Operation and Effectiveness of the Montreal Protocol's Non-Compliance Procedure', in Victor, D. G., Raustiala, K. and Skolnikoff, E. (eds) (1998a), 137–76.

Victor, D. G., Raustiala, K. and Skolnikoff, E. B. (eds) (1998a), *The Implementation and Effectiveness of International Environmental Commitments: Theory and Practice* (Cambridge, MA: The MIT Press).

Victor, D. G., Raustiala, K. and Skolnikoff, E. (1998b), 'Introduction and Overview', in Victor, D. G., Raustiala, K. and Skolnikoff, E. B. (eds), 1–46.

Vidas, D. (1996), 'The Antarctic Treaty System in the International Community: An Overview', in Stokke, O. S. and Vidas, D. (eds), 35–60.

Vidas, D. and Ostreng, W. (eds) (1999), *Order for the Oceans at the Turn of the Century* (The Hague: Springer).

Wapner, P. (1996), *Environmental Activism and World Civic Politics* (Albany: State University of New York Press).

Wapner, P. (1997), 'Governance in Global Civil Society', in Young, O. R. (ed.), 65–84.

Weber, M. (1968), *Economy and Society: An Outline of Interpretive Sociology* (New York: Bedminster Press).

Weiss, E. B. and Jacobsen, H. K. (eds) (1998), *Engaging Countries: Strengthening Compliance with International Environmental Accords* (Cambridge, MA: The MIT Press).

Weiss, E. B., Magraw, D. B. and Szasz, P. C. (1999), *International Environmental Law: Basic Instruments and References 1992–1999* (Ardsley: Transnational Publishers).

Wettestad, J. (1999), *Designing Effective Regimes: The Key Conditions* (Cheltenham: Edward Elgar).

Wettestad, J. (2002a), *Clearing the Air – European Advances in Tackling Acid Rain and Atmospheric Pollution* (Aldershot: Ashgate).

Wettestad, J. (2002b), 'The Vienna Convention and Montreal Protocol on Ozone-Layer Depletion', in Miles, E. L., Underdal, A., Andresen, S., Wettestad, J., Skjaerseth, J. B. and Carlin, E. M. (eds), 149–70.

Willetts, P. (2000), 'From "Consultative Arrangements" to "Partnership": The Changing Status of NGOs in Diplomacy at the UN', *Global Governance* 6:2, 191–212.

Winter, G. (ed.) (2006), *Multilevel Governance of Global Environmental Change: Perspectives from Science, Sociology and the Law* (Cambridge: Cambridge University Press).

Wolf, K. D. (1995), 'Was sind "nationale Interessen"? Versuch einer begrifflichen Orientierungshilfe im Spannungsfeld von Staatsräson, vergesellschafteter Außenpolitik und transnationalen Beziehungen', in Calließ, J. and Moltmann, B. (eds), 248–68.

Wolf, K. D. (ed.) (1997), *Projekt Europa im Übergang? Probleme, Modelle und Strategien des Regierens in der Europäischen Union* (Baden-Baden: Nomos).

Wolf, K. D. (2000), *Die Neue Staatsräson – Zwischenstaatliche Kooperation als Demokratieproblem in der Weltgesellschaft: Plädoyer für eine geordnete Entstaatlichung des Regierens jenseits des Staates* (Baden-Baden: Nomos).

Wolf, K. D. (2002), 'Contextualizing Normative Standards for Legitimate Governance beyond the State', in Grote, J. R. and Gbikpi, B. (eds), 35–50.

Wolf, K. D. (2008), 'Emerging Patterns of Global Governance: The New Interplay between the State, Business and Civil Society', in Scherer, A. G. and Palazzo, G. (eds), *Handbook of Research on Global Corporate Citizenship* (Cheltenham: Edward Elgar), 225–48.

World Meteorological Organization (WMO) (1988), *Conference Proceedings: The Changing Atmosphere: Implications for Global Security, Toronto, Canada 27-30 June 1988* (Geneva: WMO – No. 710).

Young, O. R. (1994), *International Governance: Protecting the Environment in a Stateless Society* (Ithaca: Cornell University Press).

Young, O. R. (ed.) (1997), *Global Governance: Drawing Insights from the Environmental Experience* (Cambridge, MA: The MIT Press).

Young, O. R. (1999a), *Governance in World Affairs* (Ithaca: Cornell University Press).

Young, O. R. (1999b), 'Regime Effectiveness: Taking Stock', in Young, O. R. (ed.), 249–79.

Young, O. R. (ed.) (1999), *The Effectiveness of International Environmental Regimes: Causal Connections and Behavioral Mechanisms* (Cambridge, MA: The MIT Press).

Young, O. R. (2001), 'Inferences and Indices: Evaluating the Effectiveness of International Environmental Regimes', *Global Environmental Politics* 1:1, 99–121.

Young, O. R. (2002a), 'Are Institutions Intervening Variables or Basic Causal Forces? Causal Clusters vs. Causal Chains in International Society', in Brecher, M. and Harvey, F. (eds), 176–91.

Young, O. R. (2002b), *The Institutional Dimensions of Environmental Change: Fit, Interplay, and Scale* (Cambridge, MA: The MIT Press).

Young, O. R. and Levy, M. A. (with the assistance of Osherenko G.) (1999), 'The Effectiveness of International Environmental Regimes', in Young, O. R. (ed.), 1–32.

Young, O. R. and Osherenko, G. (1993), 'Testing Theories of Regime Formation: Findings from a Large Collaborative Research Project', in Rittberger, V. (ed.), 223–51.

Zacher, M. (with Sutton, B. A.) (1996), *Governing Global Networks: International Regimes for Transportation and Communication* (Cambridge: Cambridge University Press).

Zangl, B. (1999), *Interessen auf zwei Ebenen. Internationale Regime in der Agrarhandels, Währungs- und Walfangpolitik* (Baden-Baden: Nomos).

Zimmermann, S. (2002), 'Frauenbewegungen, Transfer und Trans-Nationalität. Feministisches Denken und Streben im globalen und zentralosteuropäischen Kontext des 19. und frühen 20. Jahrhunderts', in Kaelble, H., Kirsch, M. and Schmidt-Gernig, A. (eds), 262–302.

Zimmerman, W., Nikitina, E. and Clem, J. (1998), 'The Soviet Union and the Russian Federation: A Natural Experiment in Environmental Compliance', in Weiss, E. B. and Jacobsen, H. K. (eds), 291–325.

Zürn, M. (1992), *Interessen und Institutionen in der internationalen Politik: Grundlegung und Anwendung des situationsstrukturellen Ansatzes* (Opladen: Leske + Budrich).

Zürn, M. (1998), *Regieren jenseits des Nationalstaats: Globalisierung und Denationalisierung als Chance* (Frankfurt/M.: Suhrkamp).

Index